Henryk Gzyl, Silvia Mayoral, and Erika Gomes-Gonçalves
Loss Data Analysis

Also of Interest

Engineering Risk Management
Thierry Meyer, Genserik Reniers, 2022
ISBN 978-3-11-066531-4, e-ISBN (PDF) 978-3-11-066533-8

Multicriteria Decision Making
Systems Modeling, Risk Assessment and Financial Analysis for Technical Projects
Timothy Havranek, Doug MacNair, 2023
ISBN 978-3-11-076564-9, e-ISBN (PDF) 978-3-11-076586-1

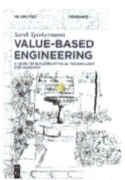
Value-Based Engineering
A Guide to Building Ethical Technology for Humanity
Sarah Spiekermann, 2023
ISBN 978-3-11-079336-9, e-ISBN (PDF) 978-3-11-079338-3

Empathic Entrepreneurial Engineering
The Missing Ingredient
David Fernandez Rivas, 2022
ISBN 978-3-11-074662-4, e-ISBN (PDF) 978-3-11-074682-2

Cloud Analytics for Industry 4.0
Edited by: Sirisha Potluri, Sachi Nandan Mohanty, Gouse Baig Mohammad, S. Shitharth, 2022
ISBN 978-3-11-077149-7, e-ISBN (PDF) 978-3-11-077157-2

Henryk Gzyl, Silvia Mayoral, and
Erika Gomes-Gonçalves

Loss Data Analysis

The Maximum Entropy Approach

2nd, extended edition

DE GRUYTER

Mathematics Subject Classification 2010
62-07, 60G07, 62G32, 62P05, 62P30, 62E99, 60E10, 60-80, 47N30, 97K80, 91B30

Authors

Prof. Dr. Henryk Gzyl
Instituto de Estudios Superiores de Administración
IESA
San Bernardino
Caracas 1010
Venezuela
henryk.gzyl@gmail.com

Dr. Erika Gomes-Gonçalves
Independent Scholar
Calle Huertas 80
Piso 2 A.
28014 Madrid
Spain
erikapat@gmail.com

Prof. Dr. Silvia Mayoral
Universidad Carlos III de Madrid
Department of Business Administration
Calle Madrid 126
28903 Getafe
Spain
silvia.mayoral@uc3m.es

ISBN 978-3-11-104738-6
e-ISBN (PDF) 978-3-11-104818-5
e-ISBN (EPUB) 978-3-11-104970-0

Library of Congress Control Number: 2022950762

Bibliographic information published by the Deutsche Nationalbibliothek
The Deutsche Nationalbibliothek lists this publication in the Deutsche Nationalbibliografie; detailed bibliographic data are available on the Internet at http://dnb.dnb.de.

© 2023 Walter de Gruyter GmbH, Berlin/Boston
Cover image: Stefan Gzyl
Typesetting: VTeX UAB, Lithuania
Printing and binding: CPI books GmbH, Leck

www.degruyter.com

On making mistakes

"The higher up you go, the more mistakes you are allowed. Right at the top, if you make enough of them, it's considered to be your style."
Fred Astaire

"A person who never made a mistake never tried anything new."
Albert Einstein

"Mistakes are the portals of discovery."
James Joyce

"Testing can show the presence of errors, but not their absence."
Edsger Dijkstra

"Makin' mistakes ain't a crime, you know. What's the use of having a reputation if you can't ruin it every now and then?"
Simone Elkeles

"If you are afraid to take a chance, take one anyway. What you don't do can create the same regrets as the mistakes you make."
Iyanla Vanzant

And to finish the list, the following is just great:

"We made too many wrong mistakes."
Yogi Berra

Preface

This is an introductory course on loss data analysis. By that we mean the determination of the density of the probability of cumulative losses. Even though the main motivations come from applications to the insurance, banking and financial industries, the mathematical problem that we shall deal with appears in many application in engineering and natural sciences, where the appropriate name would be accumulated damage data analysis or systems reliability data analysis. Our presentation will be carried out as if the focus of our attention is operational risk analysis in the banking industry.

Introductory does not mean simple: because the nature of the problems to be treated is complicated, some sophisticated tools may be required to deal with it.

Even though the main interest, which explains the title of the book, is to develop a methodology to determine probability densities of loss distributions, the final numerical problem consists of the determination of the probability density of a positive random variable. As such, the problem appears in a large variety of fields.

For example, for the risk management of a financial institution, the nature of the problem is complicated because of the very large variety of risks present. This makes it hard to properly quantify the risks as well as to establish the cause-effect relationships that may allow preventive control of the risk or the damage the risk events may cause. For risk analysts, the complication comes from the fact that the precise attribution of losses to risk events is complicated, due to the variety and nature of the risk events.

Historically speaking, in the engineering sciences there has been a large effort directed to developing methods that can identify and manage risks and quantify the damage in risk events. In parallel, in the insurance business, a similar effort has led to a collection of techniques developed to quantify damages (or losses) for the purpose of the determination of risk premia.

In the banking and financial industries, during the last few decades, there has been a collective effort to develop a precise conceptual framework in which to characterize and quantify risk, and in particular, operational risk, which basically describes losses due to the way in which business is carried out. This was done so that banks would put money aside to cover possible operational risk losses. This has resulted in a typification of risks according to a list of business types. Such typification has resulted in systematic procedures to aggregate losses in order to compute their distribution.

Similar mathematical problems also appear in systems reliability and operations research in the insurance industries; in the problem of finding the distribution of a positive random variable describing some threshold in structural engineering; and in describing the statistics of the escape time from some domain when modeling reaction rates in physical chemistry.

In all of these problems, one eventually ends up with the need to determine a probability density from the knowledge of its Laplace transform, which may be known either analytically, as the solution of some (partial) differential equation, or numerically, as the result of some simulation process or calculated from empirical data.

The problem of inverting the Laplace transform can be transformed into a fractional moments problem on the unit interval, and we shall see that the method of maximum entropy provides an efficient and robust technique to deal with these problems. This is what this book is really about.

This volume is written with several possible classes of readers in mind. First and foremost, it is written for applied mathematicians that need to address the problem of inverting the Laplace transform of probability density (or of a positive function). The methodology that we present is rather effective. Additionally, with the banking and insurance industries in mind, risk managers should be aware of the potential and effectiveness of this methodology for the determination of risk capital and the computation of premia.

To conclude, we mention that all numerical examples were produced using R.

The authors wish to thank the copy-editor and the typesetters for their professional and careful handling of the manuscript.

Madrid Erika Gomes-Gonçalves, Silvia Mayoral
Caracas Henryk Gzyl

Contents

Preface — VII

1 Introduction — 1
1.1 The basic loss aggregation problem — 1
1.2 Description of the contents — 2

2 Frequency models — 5
2.1 A short list of examples — 5
2.1.1 The Poisson distribution — 6
2.1.2 Poisson mixtures — 7
2.1.3 The negative binomial distribution — 8
2.1.4 Binomial models — 9
2.2 Unified version — 9
2.3 Examples — 10
2.3.1 Determining the parameter of a Poisson distribution — 11
2.3.2 Binomial distribution — 18

3 Individual severity models — 22
3.1 A short catalog of distributions — 22
3.1.1 Exponential distribution — 22
3.1.2 The simple Pareto distribution — 23
3.1.3 Gamma distribution — 23
3.1.4 The lognormal distribution — 24
3.1.5 The beta distribution — 24
3.1.6 A mixture of distributions — 25
3.2 Numerical examples — 25
3.2.1 Data from a density — 26

4 Some detailed examples — 30
4.1 Claim distribution and operational risk losses — 30
4.2 Simple model of credit risk — 31
4.3 Shock and damage models — 35
4.4 Barrier crossing times — 35
4.5 Applications in reliability theory — 36

5 Some traditional approaches to the aggregation problem — 38
5.1 General remarks — 38
5.1.1 General issues — 39
5.2 Analytical techniques: The moment generating function — 40
5.2.1 Simple examples — 41

5.3	Approximate methods —— 44
5.3.1	The case of computable convolutions —— 44
5.3.2	Simple approximations to the total loss distribution —— 45
5.3.3	Edgeworth approximation —— 46
5.4	Numerical techniques —— 48
5.4.1	Calculations starting from empirical or simulated data —— 48
5.4.2	Recurrence relations —— 49
5.4.3	Further numerical issues —— 50
5.5	Numerical examples —— 50
5.6	Concluding remarks —— 60
6	**Laplace transforms and fractional moment problems —— 62**
6.1	Mathematical properties of the Laplace transform and its inverse —— 62
6.2	Inversion of the Laplace transform —— 64
6.3	Laplace transform and fractional moments —— 66
6.4	Unique determination of a density from its fractional moments —— 67
6.5	The Laplace transform of compound random variables —— 67
6.5.1	The use of generating functions and Laplace transforms —— 68
6.5.2	Examples —— 70
6.6	Numerical determination of the Laplace transform —— 72
6.6.1	Recovering a density on [0, 1] from its integer moments —— 73
6.6.2	Computation of (6.5) by Fourier summation —— 74
7	**The standard maximum entropy method —— 76**
7.1	The generalized moment problem —— 77
7.2	The entropy function —— 78
7.3	The maximum entropy method —— 80
7.4	Two concrete models —— 82
7.4.1	Case 1: The fractional moment problem —— 82
7.4.2	Case 2: Generic case —— 82
7.5	Numerical examples —— 83
7.5.1	Density reconstruction from empirical data —— 83
7.5.2	Reconstruction of densities at several levels of data aggregation —— 87
7.5.3	A comparison of methods —— 95
8	**Extensions of the method of maximum entropy —— 97**
8.1	Generalized moment problem with errors in the data —— 98
8.1.1	The bounded error case —— 98
8.1.2	The unbounded error case —— 100
8.1.3	The fractional moment problem with bounded measurement error —— 101
8.2	Generalized moment problem with data in ranges —— 102
8.2.1	Fractional moment problem with data in ranges —— 104

8.3	Generalized moment problem with errors in the data and data in ranges —— 105
8.4	Numerical examples —— 106
8.4.1	Reconstruction from data in intervals —— 107
8.4.2	Reconstruction with errors in the data —— 109

9 Superresolution in maxentropic Laplace transform inversion —— 113
9.1	Introductory remarks —— 113
9.2	Properties of the maxentropic solution —— 114
9.3	The superresolution phenomenon —— 115
9.4	Numerical example —— 117

10 Sample data dependence —— 119
10.1	Preliminaries —— 119
10.1.1	The maximum entropy inversion technique —— 120
10.1.2	The dependence of λ on μ —— 122
10.2	Variability of the reconstructions —— 123
10.2.1	Variability of expected values —— 127
10.3	Numerical examples —— 128
10.3.1	The sample generation process —— 128
10.3.2	The 'true' maxentropic density —— 129
10.3.3	The sample dependence of the maxentropic densities —— 130
10.3.4	Computation of the regulatory capital —— 134

11 Disentangling frequencies and decompounding losses —— 136
11.1	Disentangling the frequencies —— 136
11.1.1	Example: Mixtures of Poisson distributions —— 138
11.1.2	A mixture of negative binomials —— 143
11.1.3	A more elaborate case —— 143
11.2	Decompounding the losses —— 146
11.2.1	Simple case: A mixture of two populations —— 148
11.2.2	Case 2: Several loss frequencies with different individual loss distributions —— 150

12 Computations using the maxentropic density —— 152
12.1	Preliminary computations —— 152
12.1.1	Calculating quantiles of compound variables —— 153
12.1.2	Calculating expected losses given that they are large —— 154
12.1.3	Computing the quantiles and tail expectations —— 154
12.2	Computation of the VaR and TVaR risk measures —— 155
12.2.1	Simple case: A comparison test —— 155
12.2.2	VAR and TVaR of aggregate losses —— 158

12.3 Computation of risk premia — **159**

13 **A solution to the capital allocation problem — 162**
13.1 Introduction and preliminaries — **162**
13.2 Numerical results — **164**
13.2.1 The capital allocation — **164**
13.2.2 Problem 2: Determining the distortion function from given risk prices — **166**
13.3 Appendices — **172**
13.3.1 Application of MEM to determine the capital allocation — **172**
13.3.2 Distorted risk measure — **175**
13.3.3 Application of MEM to determine the distortion measure — **176**
13.3.4 A numerical instability issue — **177**

14 **Review of statistical procedures — 179**
14.1 Parameter estimation techniques — **179**
14.1.1 Maximum likelihood estimation — **179**
14.1.2 Method of moments — **180**
14.2 Clustering methods — **181**
14.2.1 K-means — **181**
14.2.2 EM algorithm — **182**
14.2.3 EM algorithm for linear and nonlinear patterns — **183**
14.2.4 Exploratory projection pursuit techniques — **185**
14.3 Approaches to select the number of clusters — **187**
14.3.1 Elbow method — **187**
14.3.2 Information criteria approach — **187**
14.3.3 Negentropy — **188**
14.4 Validation methods for density estimations — **188**
14.4.1 Goodness of fit test for discrete variables — **189**
14.4.2 Goodness of fit test for continuous variables — **189**
14.4.3 A note about goodness of fit tests — **191**
14.4.4 Visual comparisons — **192**
14.4.5 Error measurement — **192**
14.5 Beware of overfitting — **193**
14.6 Copulas — **194**
14.6.1 Examples of copulas — **195**
14.6.2 Simulation with copulas — **202**

Bibliography — 205

Index — 209

1 Introduction

In this short chapter we shall do two things. First, we briefly describe the generic problem that eventually leads to the same typical numerical Laplace inversion problem, and second, we give a summary of the contents of this book.

1.1 The basic loss aggregation problem

Consider the following compound random variable:

$$S = \sum_{n=1}^{N} X_n, \tag{1.1}$$

in which N is an integer valued random variable and the X_n are supposed to be positive random variables, independent identically distributed and independent of N, all of which are defined on a common probability space $(\Omega, \mathcal{F}, P)$. The product space structure from the underlying probability space comes to mind as the obvious choice. That is, we may consider $\Omega = \mathbb{N} \times [0, \infty)^{\mathbb{N}}$, with typical sample point $\omega = (n, x_1, x_2, \dots)$ and coordinate maps $N: \Omega \to \mathbb{N}$ defined by $N(\omega) = n$, and $X_n(\omega) = x_n$. All questions that we can ask about this model are contained in the σ-algebra $\mathcal{F}$ generated by N and the X_n.

Here, we suppose that the distributions $p_n = P(N = n), n = 0, 1, \dots$ and $F(x) = P(X_n \leq x), x \in (0, \infty)$ are supposed to be known, and the problem that we are interested in solving is to determine $P(S \leq x), x \in (0, \infty)$. Since in many of the applications of interest to us the X_n have a density, the problem we want to solve is to find the probability density of S from the information available about N and the X_n, or as may be the case in many applications, from the empirical knowledge of S.

In the list below we mention a few concrete examples. These will be further explained and amplified in Chapter 4.
- In the insurance industry: N is the total number of claims in a given time period, and X_n is the amount claimed by the n-th claimant.
- In the banking industry: N stands for the number of risk events of a certain type, say credit card frauds or typing mistakes authorizing a money transfer at certain branch of the bank. In this case X_n may stand for the loss in each theft or the cost of each typing mistake. In this case, S will stand for the total loss due to each risk type.
- Again, in the banking industry, consider the following simplified model of credit risk. At the beginning of the year, the bank concedes a certain number of loans, some of which are not paid by the end of the year. Now, N stands for the total number of customers that do not return a loan at the end of the year. In this case N is a random number between 0 and the total number of loans that the bank has given, which are due within the period. This time X_n is the amount not returned by the n-th borrower. If all borrowers are supposed to be statistically equivalent, we are within the framework of our basic model.

– In operations research: Let N be the number of customers taken care of at a certain business unit, and X_n the total money that each of them spends, or the time that it takes to serve each of them. Then, S will denote either the total receipt for the period or the total service time, and the management might be interested in its probability distribution.

Some of these problems lead to subproblems that are interesting in themselves, and important to solve as part of the general problem at hand. Consider for example the following situation. Suppose that we record global losses that result from the aggregation of some other, more basic losses. Furthermore, suppose that the number of possible risk sources is also known. A question of interest is the following: Can anything be said about the individual loss frequencies and the individual severities at lower levels of aggregation? We mention two specific examples. An insurance company suspects that the frequency of claims and cost of the claims incurred by male or female claimants are different, and wants to know by how much. Or consider typing errors at a bank, which when made by male or female clerks may cause different losses. These examples pose the problem of determining what kind of individual information can be extracted from the aggregate data.

It is one of the purposes of this book to propose a methodology to solve the various problems that appear when trying to determine the probability distribution of S from the distribution of N and that of the X_n, and then to determine the probability distribution resulting from aggregating such collective risks. This is an old problem, which has been addressed in many references. The other aim is to examine some related questions, like how the methodology that we propose depend on the available data, or why it works at all.

1.2 Description of the contents

The initial chapters follow a more or less traditional approach, except for the fact that we do not review the very basic notions of probability, calculus and analysis. We start the book by devoting Chapter 2 to describing some of the integer valued random variables that are used to model the frequency of events. As we shall be considering events that take place during a fixed interval of time (think of a year, to be specific), we will not be considering integer valued random processes, just plain integer valued random variables.

In Chapter 3 we consider a variety of the standard examples to describe individual severities. The reason for considering this in two blocks was mentioned previously. We want to obtain the statistical properties of the compound variable S from the properties of its building blocks. Before reviewing some of the traditional ways of doing that in Chapter 5, we devote Chapter 4 to looking in more detail at the examples mentioned after (1.1), that is, we will describe several possible problems leading to the need to determine the statistical nature of a compound random variable.

This is a good point to say that readers familiar with the contents of Chapters 2, 3, and 5 (take a glimpse at the table of contents), can skip the first five chapters and proceed directly to Chapter 6. These chapters are meant to fill in some basic vocabulary in probability theory for professionals from other disciplines, so the material is included for completeness sake.

Chapter 6 is devoted to the Laplace inversion problem, because many traditional approaches rely on it. From the knowledge of a model for the frequency of events N and a model for the individual loss distribution losses X, the Laplace transform on the aggregate loss can be simply computed if the independence assumption is brought in. Then the problem of determining the density of the aggregate loss becomes a problem of inverting a Laplace transform. That is why we review many of the traditional methods for inverting Laplace transforms.

The methodology that we shall develop to invert Laplace transforms to the density of S relies on the connection between the problem of inverting the Laplace transform of a positive random variable and the fractional moment problem. As we shall see, the applicability of the methodology that we propose goes beyond that of finding the density of compound random variables. We shall describe that connection in Section 6.4.

In Chapters 7 and 8 we consider the basic aspects of mathematical techniques for solving the fractional moment problem, namely the standard method of maximum entropy. This method is useful for several reasons. First, it does not depend on the knowledge of underlying models and therefore it is nonparametric. Second, it requires little data. In fact, eight fractional moments (or the values of the Laplace transform at eight points) lead to quite good reconstructions of the density. We explain why that might be so in Chapter 9. Last but not least, the method allows for extensions to cases in which the data is known to fall in ranges instead of being point data, or extensions to incorporate error measurements in the available data.

Since the reconstructions fit the data rather well, that is the Laplace transform of the maxentropic density coincides with the given values of the Laplace transform regardless of how the transform was calculated, we can easily study how the transform depends on the available data. We examine this issue both theoretically and by means of simulations in Chapter 10.

In Chapter 11 we address a natural converse problem: If we are given aggregate data, besides being able to determine the probability density of the aggregate distribution, we shall see that to some extent, we can determine the underlying statistical nature of the frequency of events and that of the individual losses. In Chapter 12 we consider some standard applications, one of the more important being the computation of some risk measures, like value at risk (VaR) and tail value at risk (TVaR).

Chapter 13 is devoted to an application of the method of maximum entropy in the mean to a related risk management problem. After an enterprise has determined the risks in each line of activity, it has to decide how much rik capital to allocate to cover each possible risk when the total risk capital is preassigned.

To finish, in Chapter 14 we review some of the statistical procedures used throughout the book. In particular, we emphasize the statistical procedures used to check the quality of the maxentropically reconstructed density functions, and review the notion of copula, which is used to produce joint distributions from individual distributions.

2 Frequency models

One of the two building blocks that make up the simplest model of aggregate severity is the frequency of events during the modeling period. We shall examine here some of the most common examples of (positive) integer valued random variables and some of their properties. An integer valued random variable is characterized by its probability distribution $P(N = k) = p_k;\ k = 0, 1, \ldots$. Whether we consider bounded or unbounded random variables depends on the application at hand: If we consider the random number N of defaults among a 'small' number M of borrowers, it is reasonable to say that $0 \leq N \leq M$. But if M is a large number, it may be convenient to suppose that N can be modeled by an unbounded random variable.

To synthesize properties of the collection $\{p_k, k \geq 0\}$ and to study statistical properties of collections of random variables, it is convenient to make use of the concept of the generating function. It is defined by:

Definition 2.1. Consider an integer valued random variable, and let z be a real or complex number with $|z| < 1$. The series

$$G_N(z) = \sum_{k \geq 0} p_k z^k = \mathrm{E}[z^N]$$

is called the generating function of the sequence $\{p_k, k \geq 0\}$.

In many of the models used to describe the frequency of events, there exists a relationship among every p_k that makes it possible to compute explicitly the generating function $G(z)$. Clearly, the series defining $G(z)$ converges and $(1/k!) d^k G/dz^k(0) = p_k$. One of the main applications of the concept comes through the following simple lemma:

Lemma 2.1. *Suppose that $N_1, \ldots, N_K$ are independent integer valued random variables. With the notation just introduced, if $N = N_1 + N_2 + \cdots + N_K$, then*

$$G_N(z) = \prod_{j=1}^{K} G_{N_j}(z).$$

The proof is simple: Just note that $G_N(z) = \mathrm{E}[z^N] = \prod_j \mathrm{E}[z^{N_j}] = \prod_j G_{N_j}(z)$. We leave it for the reader to verify that:

Lemma 2.2. *With the notations introduced above, the moments of N are*

$$\mathrm{E}[N^k] = \left(z\frac{d}{dz}\right)^k G_N(z)\bigg|_{z=1}.$$

2.1 A short list of examples

We shall briefly describe some common examples. We will make use of them later on.

2.1.1 The Poisson distribution

This is one of the most common models. It has interesting aggregation properties. We say that N is a Poisson random variable with parameter λ if

$$P(N = k) = p_k = \frac{\lambda e^{-\lambda}}{k!}; \quad k \geq 0. \tag{2.1}$$

It takes an easy computation to verify that

$$G_N(z) = E[z^N] = e^{\lambda(z-1)}$$

is actually defined for all values of z. From this it is simple to verify that

$$E[N] = \lambda; \quad \text{Var}(N) = \lambda.$$

That is, estimating the mean of a Poisson random variable amounts to estimating its variance.

Suppose that we are interested in the frequency of events in a business line at a bank. Suppose as well that the different risk types have frequencies of Poisson type. Then the collective or aggregate risk in that business line is Poisson. Formally,

Theorem 2.1. *If $N_1, N_2, \ldots, N_K$ are Poisson with parameters $\lambda_1, \lambda_2, \ldots, \lambda_K$, then $N = N_1 + N_2 + \cdots + N_K$ is Poisson with parameter $\lambda = \sum_{i=1}^{K} \lambda_i$.*

Proof. According to Lemma 2.1,

$$G(z) = E[z^N] = E[z^{\sum N_i}] = E\left[\prod z^{N_i}\right] = \prod E[z^{N_i}]$$
$$= \prod e^{\lambda_i(z-1)} = e^{\sum \lambda_i(z-1)} = e^{\lambda(z-1)}.$$

As the generating function determines the distribution, the conclusion emerges. □

What about the opposite situation? That is, suppose that you have recorded the annual frequency of a certain type of event resulting from the aggregation of independent events, each of which occurs with known probability. For example, car crashes are the result of aggregating collisions caused by either male or female drivers, each proportional to their frequency in the population; or credit card fraud according to some idiosyncratic characteristic. What can be said about the frequency within each group from the global observed frequency? A possible way to answer this question is contained in the following result.

Theorem 2.2. *Suppose that N is Poisson with parameter λ. Suppose that each event can occur as one of K possible types that occur independently of each other with probabilities $p_1, \ldots, p_K$. The number N_j of events of type $j = 1, \ldots, K$ can be described by a Poisson frequency of parameter $\lambda_j = p_j \lambda$ and $N = \sum_{j=1}^{K} N_j$.*

Proof. Given that the event $\{N = n\}$ occurs, the individual underlying events occur according to a possibility like $\{N_1 = n_1, \ldots, N_K = n_K\}$. As the individual events in the curly brackets are independent with probabilities $p_1, \ldots, p_K$, the joint probability of that event is

$$P(N_1 = n_1, \ldots, N_K = n_K \mid N = n) = \binom{n}{n_1, \ldots, n_k} p_1^{n_1} \cdots p_K^{n_K}.$$

Therefore, since $\sum n_k = n$ and $P(N = n) = e^{-\lambda} \lambda^n / n!$,

$$P(N_1 = n_1, \ldots, N_K = n_K) = P(N_1 = n_1, \ldots, N_K = n_K \mid N = n) P(N = n)$$

$$= \binom{n}{n_1, \ldots, n_k} p_1^{n_1} \cdots p_K^{n_K} \frac{\lambda^n}{n!} e^{-\lambda}.$$

Therefore, since $\sum n_k = n$ and $\sum p_k = 1$, we have $\lambda^n = \prod_j \lambda^{n_j}$ and $e^{-\lambda} = \prod_j e^{-p_j \lambda}$. In other words, we can rewrite the last identity as

$$P(N_1 = n_1, \ldots, N_K = n_K) = \prod_{j=1}^{K} \frac{(p_j \lambda)^{n_j} e^{-p_j \lambda}}{n_j!},$$

which is equivalent to saying that we can regard N as a sum of independent Poisson random variables with parameters $\lambda_j = p_j \lambda$. □

2.1.2 Poisson mixtures

Consider the following interesting situation. Suppose that it is reasonable to model the frequency of some event by means of a Poisson distribution, but that regretfully its frequency is unknown, and it may itself be considered to be a random variable. The risk event is observed, but its source (which determines its frequency) is inaccessible to observation. To model this situation we proceed as follows:

We suppose that the values of the parameter determining the frequency are modeled by a random variable Λ, and that when the event takes place $\{\Lambda = \lambda\}$, the probability law of N is known to be $P(N = k \mid \lambda)$. That is, to be really proper we should write $P(N = k \mid \Lambda = \lambda)$, which we furthermore suppose to be given by (2.1). For the model to be complete, we should provide the distribution function $P(\Lambda \leq \lambda) = F_\Lambda(\lambda)$ of the underlying variable. Once this is specified, we can obtain the unconditional probability distribution of N by

$$P(N = k) = \int P(N = k \mid \lambda) dF_\Lambda(\lambda).$$

Consider the following examples.

Example 1. Suppose that Λ is distributed according to a $\Gamma(r,\beta)$ law. That is, Λ has a probability density given by $f_\Lambda(\lambda) = \frac{\lambda^{r-1}e^{-\lambda/\beta}}{\beta^r \Gamma(r)}$ where, as usual, $\Gamma(r)$ denotes the Euler gamma function. Then,

$$P(N=k) = \int_0^\infty P(N=k \mid \lambda) f_\Lambda(\lambda) d\lambda = \int_0^\infty \frac{\lambda^k e^{-\lambda}}{k!} \frac{\lambda^{r-1} e^{-\lambda/\beta}}{\beta^r \Gamma(r)} d\lambda$$

$$= \frac{1}{k!} \frac{1}{\beta^r \Gamma(r)} \int_0^\infty e^{-\lambda(1+\beta^{-1})} \lambda^{k+r-1} d\lambda = \frac{\Gamma(k+r)\beta^k}{k!\Gamma(r)(1+\beta)^{(k+a)}}$$

$$= \binom{k+r-1}{k} \left(\frac{\beta}{1+\beta}\right)^k \left(\frac{1}{1+\beta}\right)^r.$$

Example 2. Let us suppose that Λ is a binary random variable. Suppose that there are two types of drivers, of 'aggressive' and 'defensive' types. The insurance company only knows the statistical frequency with which they exist in the population of drivers. Or think perhaps about property damage, which may be 'intentional' or 'unintentional'. Let us model this by saying that $P(\Lambda = \lambda_1) = p$ and $P(\Lambda = \lambda_0) = 1 - p$ describes the probability distribution of Λ. Therefore, in a self explanatory notation

$$P(N = k) = P(N = k \mid \lambda_0) P(\Lambda = \lambda_0) + P(N = k \mid \lambda_1) P(\Lambda = \lambda_1),$$

or equivalently

$$P(N = k) = (1-p) \frac{\lambda_0^k}{k!} e^{-\lambda_0} + p \frac{\lambda_1^k}{k!} e^{-\lambda_1}.$$

Of course, the remaining problem is to determine p, λ_0 and λ_1 from the data.

2.1.3 The negative binomial distribution

A few lines above we explained how the following distribution appears as a mixture of Poisson random variables when the intensity is distributed according to a gamma law. An integer valued random variable is said to have a negative binomial distribution with parameters $r > 0$ and $\beta > 0$ whenever

$$P(N = k) = p_k = \binom{k+r-1}{k} \left(\frac{1}{1+\beta}\right)^r \left(\frac{\beta}{1+\beta}\right)^k. \tag{2.2}$$

It is not difficult to verify that the generating function of this probability law is

$$G(z) = \sum_{n \geq 0} p_n z^n = [1 - \beta(z-1)]^{-r},$$

which converges for $|z| < 1$. Using this result, one can verify that

$$E[N] = r\beta; \quad \text{Var}(N) = r\beta(1 + \beta),$$

which relates the parameter r and β to standard statistical moments.

Note that the *geometrical distribution* is a particular case of the negative binomial corresponding to setting $r = 1$. In this case we obtain

$$P(N = k) = p_k = \frac{1}{1+\beta}\left(\frac{\beta}{1+\beta}\right)^k = pq^k, \tag{2.3}$$

where at the last step we set $p = 1/(1 + \beta)$ and $q = 1 - p$. If we think of N as describing the number of independent trials (at something) before the first success, clearly N will have such a law. Note as well that

$$P(N \geq k) = q^k, \quad \text{and} \quad P(N \geq m + k \mid N \geq k) = q^m = P(N \geq m).$$

It is also interesting to note that when r is a positive integer and p_k is given by (2.2), then N can be interpreted as the r-th event occurring for the first time at $k + r$ attempts.

2.1.4 Binomial models

This is the simplest model to invoke when one knows that a maximum number n of independent events can occur within the period, all with the same probability of occurrence. In this case probability of occurrence of k events is clearly given by

$$P(N = k) = \binom{n}{k} p^k (1 - p)^{n-k}. \tag{2.4}$$

The generating function of this law is given by $G(z) = [1 - p(z - 1)]^n$, from which it can easily be obtained that

$$E[N] = pn; \quad \text{Var}(N) = np(1 - p).$$

Now, there is only one parameter to be estimated, and it is related to the mean.

2.2 Unified version

There is an interesting way of describing the four families of variables that we have discussed above. The reason is more than theoretical, because the characterization provides an inference procedure complementary to those described above, in which we related the moments of the variables to the parameters of their distributions.

Definition 2.2. Denote by p_k the probability $P(N = k)$. We shall say that it belongs to the Panjer $(a, b, 0)$ class if there exist constants a, b such that

$$p_k/p_{k-1} = a + b/k, \quad \text{for } k = 1, 2, 3, \ldots. \tag{2.5}$$

Note that the value of p_0 has to be adjusted in such a way that $\sum_{n \geq 0} p_n = 1$.

The relationship between a, b and the standard parameters is described in Table 2.1.

Table 2.1: The Panjer $(a, b, 0)$ class.

Distribution	a	b	p_0
Poisson	0	λ	$e^{-\lambda}$
Binomial	$-\frac{p}{1-p}$	$(n+1)\frac{p}{1-p}$	$(1-p)^n$
Neg. Binomial	$\frac{\beta}{1+\beta}$	$(r-1)\frac{\beta}{1+\beta}$	$(1+\beta)^{-r}$
Geometric	$\frac{\beta}{1+\beta}$	0	$(1+\beta)^{-r}$

The interesting feature of the characterization mentioned above is the following: Suppose that we have a 'reasonable amount' N of observations so that we can estimate $\hat{p}_k = \frac{n_k}{n}$, where n_k is the number of time periods in which exactly k risk events occurred; then (2.5) can be rewritten as

$$k\hat{p}_k/\hat{p}_{k-1} = ka + b; \quad k = 1, 2, \ldots.$$

The beauty about this is that it can be reinterpreted as linear regression data, from which the parameter a, b can be estimated and the standard parameters of the distribution can be obtained as indicated in Table 2.1.

2.3 Examples

In this section we examine in more detail the process of fitting a model to empirical data about the frequency of events of some (risk) type. The usual first choices that come to mind are the binomial, negative-binomial and Poisson models. We mention at this point that we shall make extensive use of this class in Chapter 12 when studying the disentangling problem. For the time being we shall illustrate the most direct applications of this class.

2.3.1 Determining the parameter of a Poisson distribution

Suppose we have data about daily system failures at a given bank, and that we have organized the data as shown in Table 2.2.

Table 2.2: Daily system failures of some bank.

Number of events (k)	Frequency (n_k)	Relative frequency (p_k)
0	11	0.011
1	50	0.050
2	112	0.112
3	169	0.169
4	190	0.190
5	171	0.171
6	128	0.128
7	82	0.082
8	46	0.046
9	23	0.023
10	10	0.010
11	5	0.005
12	2	0.002
13	1	0.001
14	0	0.000
Total	1000	1

In the left column of Table 2.2 we list the possible number of failures and in the second column we list the number of days in which that number of failures occurred. Clearly, during the data gathering period, more than 14 failures were never observed during one day. The histogram corresponding to that data is displayed in Figure 2.1.

The first step in the risk modeling process is to determine a model for the frequency of losses. So, let us suppose that the frequency model belongs to the $(a, b, 0)$ family. Thus, we shall consider the characterization

$$k \frac{\hat{p}_k}{\hat{p}_{k-1}} = ak + b,$$

where, to be explicit, $\hat{p}_k = \frac{n_k}{\sum_k n_k}$, and n_k is the frequency with which k losses are observed.

Let us now consider Table 2.3, a variant of the first table in which the last column is replaced by a column displaying $k \frac{\hat{p}_k}{\hat{p}_{k-1}} = k \frac{n_k}{n_{k-1}}$.

Next, we plot $k \frac{n_k}{n_{k-1}}$ versus k to obtain Figure 2.2. Clearly the values appear to be constant about 4.5 except for the three last values. This would suggest that we are in

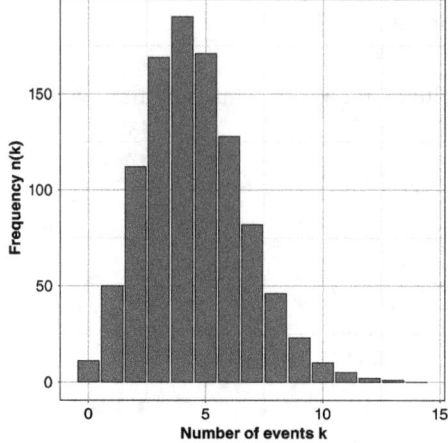

Figure 2.1: Failures versus time.

Table 2.3: Daily system failures in some bank.

Number of events (k)	Frequency (n_k)	$k \frac{n_k}{n_{k-1}}$
0	11	–
1	50	4.545
2	112	4.480
3	169	4.527
4	190	4.497
5	171	4.500
6	128	4.491
7	82	4.484
8	46	4.488
9	23	4.500
10	10	4.348
11	5	5.500
12	2	4.800
13	1	6.500
14	0	0.000
Total	1000	–

the presence of a Poisson distribution, because a appears to be 0 and b (which for the Poisson coincides with λ) seems to be about 4.5.

One easy test of the 'Poissonness' of the distribution is the comparison of the mean and the variance of the data. Both the mean and the variance of the data in Table 2.2 equal 4.5. This reinforces the possibility of its being a Poisson distribution. To be really sure, we would need to apply tests like the chi-square (χ^2), maximum likelihood or Kolmogorov–Smirnov tests, in case we were dealing with continuous data.

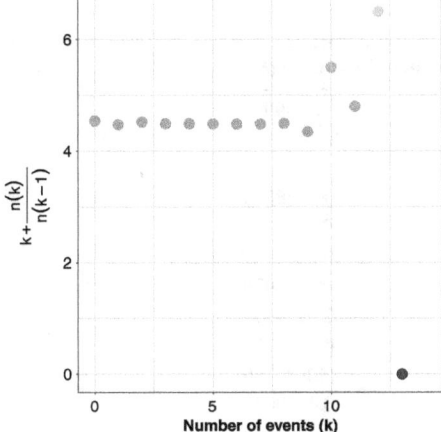

Figure 2.2: Plot of k versus $k\frac{n_k}{n_{k-1}}$.

χ^2 goodness of fit test

This is a standard hypothesis test. Here we shall apply it to the data in Table 2.2 to decide whether the data comes from a Poisson of parameter $\lambda = 4.5$. The first step consists of establishing the null and alternative hypotheses:
(1) $\mathbf{H_0}$: the random variable is distributed according to a Poisson law
(2) $\mathbf{H_1}$: the random variable is not distributed according to a Poisson law.

In Figure 2.1 we display the frequency histogram of the daily data. The number of categories (bins) is $k = 14$ and the total number of data points is $n = 1000$ (roughly 40 months of data) and there is only one parameter (λ) to be estimated. We already did that, but we can also obtain it from the estimated sample mean. The observed frequency O_i in each of the bins $\{0, 1, \ldots, 14\}$ is shown in Table 2.2. The expected frequency E_i for each of the k categories can be estimated by $E_i = nP(X = i) = n(e^{-\lambda}\lambda^i/i!)$, with $\lambda = 4.5$. In Table 2.4 we gather all the steps necessary to compute the value of χ^2_o. As a practical rule, it is advised that the theoretical expected frequency in each category is not less than 5. In case it is to be applied, one may combine successive intervals until the minimal frequency is achieved.

In Table 2.4 we see that the practical rule suggests that we should combine intervals to form a new category. In our case, this will correspond to 11 or more events. In Table 2.5 we see the resulting new dataset, in which the number of categories is 12. Therefore the new degree of freedom for the χ^2 test is $12 - 1 - 1 = 10$. If we set the level of significance to $\alpha = 0.05$, the critical value of $\chi(0.05, 10)$ can be obtained from the tables of the χ^2 distribution or using your favorite software.

Therefore, $\chi^2_o = \sum_{i=1}^{k} \frac{(O_i - E_i)^2}{E_i} = 0.2922061$. The value of $\chi^2_{0.05,10} = 18.307045$ using the R command $qchisq(1 - 0.05, 10)$. Since $0.2922061 < 18.30704$ we do not reject the null hypothesis and we conclude that the data is distributed according to a Poisson(4.5) law.

Table 2.4: Absolute and relative frequencies in each category.

Number of events (k)	Absolute frequency (O_k)	Relative frequency (E_k)
0	11	11.1089965
1	50	49.9904844
2	112	112.4785899
3	169	168.7178849
4	190	189.8076205
5	171	170.8268585
6	128	128.1201439
7	82	82.3629496
8	46	46.3291592
9	23	23.1645796
10	10	10.4240608
11	5	4.2643885
12	2	1.5991457
13 or more	1	0.8051379
Total	1000	1000

Table 2.5: Absolute and expected frequencies.

Number of events (k)	Absolute frequency (O_k)	Expected frequency (E_k)
0	11	11.1089965
1	50	49.9904844
2	112	112.4785899
3	169	168.7178849
4	190	189.8076205
5	171	170.8268585
6	128	128.1201439
7	82	82.3629496
8	46	46.3291592
9	23	23.1645796
10	10	10.4240608
11 or more	8	6.6686721
Total	1000	1000

The procedure just described can be carried out in R using the command *goodfit* available in the *vcd* library. This command allows us to verify whether a dataset does or does not fit some of the discrete distributions (binomial, negative binomial or Poisson) via a χ^2 or a maximum likelihood procedure.

Organizing the data in a matrix x, the code for the R procedure for the goodness of fit test is displayed in Table 2.6.

The results of applying the code in Table 2.6 are displayed in Table 2.7. As we mentioned above, the dataset in Table 2.2 fits a Poisson quite well.

Table 2.6: R code for the goodness of fit test.

```
# the original data is grouped, the following instructions ungroup them:
xpoi=NULL
for (k in min(x[,1]):max(x[,1]))
        xpoi=c(xpoi,rep(k,x[k+1,2]))
xpoi
#apply the χ² test:
library(vcd)
gf=goodfit(xpoi,type= "poisson",method= "MinChisq")
summary(gf)
plot(gf,main="Frequency vs. Poisson dist.",xlab = 'Number of failures', ylab = 'days')
```

Table 2.7: Results of applying the goodness of fit test.

Goodness of fit test for Poisson distribution		
X^2	df	$P(> X^2)$
Pearson 0.291703	12	1
In summary.goodfit(gf) : Chi-squared approximation may be incorrect		

Comment. Note that the number of degrees of freedom in Table 2.7 is different from its true value because the procedure 'does not know' that we are estimating a parameter. We mention as well that in R there exists another command that produces similar results, namely the command *chisq.test*.

A simpler but not so rigorous numerical procedure consists of plotting the probability distribution that is supposed to describe the data along with a histogram of the data, and comparing the result visually. For the dataset in Table 2.5, such a procedure yields Figure 2.3. The display suggests that the proposed distribution seems to be the right one.

Computation of the *P*-value

The *P*-value is usually used in hypothesis testing. It is defined as the smallest significance level that can be chosen to possibly reject the null hypothesis about a given dataset. Any significance level smaller than the *P*-value implies not rejecting the null hypothesis.

In the present case it can be defined by:

$$P_{\text{value}} \approx P(\chi^2_{a,k-p-1} > \chi^2_0)$$

$$0 \leq P_{\text{value}} \leq 1.$$

The null hypothesis is rejected when:

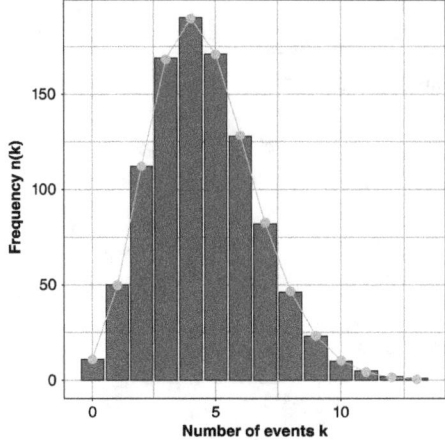

Figure 2.3: Frequency data and theoretical distribution.

$$\alpha > P_{\text{value}}.$$

Thus, in our example it happens that

$$P_{\text{value}} \approx P(\chi^2_{0.5,10} > 0.2922061) = 0.9999995.$$

Observe that the results using R, displayed in Table 2.7, show that the P_{value} is approximately 1, which coincides with the previously obtained result. The P_{value} can be computed using R by $1 - pchisq(0.29, 10)$.

Comment. This test is independent of the sample size, so it is always possible to reject a null hypothesis with a large enough sample, even though the true difference is very small. This situation can be avoided by complementing the analysis with the calculation of the power of the test and the optimal sample space, as suggested, for example, in [2].

Determination of the parameter by maximum likelihood

The basic philosophy of the maximum likelihood estimation procedure is the following: We suppose that the distribution belongs to some parametric distribution family, and then we determine the parameter that makes the dataset most probable. Then, the goodness of fit test may be applied to complement the analysis.

So, consider the data in Table 2.2 once more. We begin by supposing that the distribution underlying the data is Poisson $P(X = k) = \frac{e^{-\theta}\theta^k}{k!}$, and the likelihood function is defined by

$$L(\theta; k_1, k_2, \ldots, k_n) = \prod_{i=1}^{n} f(k_i \mid \theta) = \prod_{i=1}^{n} \frac{e^{-\theta}\theta^{k_i}}{k_i!}$$

$$= \frac{e^{-n\theta}\theta^{\sum_{i=1}^{n} k_i}}{\prod_{i=1}^{n} k_i!}.$$

In our example $n = 1000$. The log-likelihood function is defined by

$$\log L(\theta; x_1, x_2, \ldots, x_n) = -n\theta + \log \theta \sum_{i=1}^{n} x_i - \log\left(\prod_{i=1}^{n} x_i!\right),$$

where the term $\log(\prod_{i=1}^{n} x_i!)$ can be ignored because the likelihood function is maximized with respect to θ. To find the value of θ that maximizes the likelihood, we differentiate the log-likelihood function with respect to θ and equate the result to zero, obtaining

$$\frac{\partial \log L(\theta; x_1, x_2, \ldots, x_n)}{\partial \theta} = -n + \frac{\sum_{i=1}^{n} x_i}{\theta} = 0.$$

Therefore,

$$\hat{\theta} = \frac{\sum_{i=1}^{n} x_i}{n},$$

which happens to be the sample mean of the data, with a value of 4.505. To verify that $\hat{\theta}$ really minimizes the likelihood function, we compute its second derivative and substitute in the value of $\hat{\theta}$, obtaining

$$\frac{\partial^2 \log L(\theta; x_1, x_2, \ldots, x_n)}{\partial^2 \theta} = -\frac{\sum_{i=1}^{n} x_i}{\theta^2} = -\frac{n^2}{\sum_{i=1}^{n} x_i} < 0.$$

If you are a fan of statistical software, the procedure just described can be carried out using R as described in Table 2.8. The user information about this, or any other command, may be obtained by invoking the "help"(command).

Application of the method of moments

The method of moments is a useful technique on two counts. First, some probability distributions are determined by the collection of their (integer) moments, and second, for many parametric distributions the parameters are related to the moments in a simple way.

From the analytic point of view there is a simple relationship between the moments and the generating function of the probability distribution. Let us work out some details in the context of a simple example.

Suppose that we wanted to determine the parameter λ of a Poisson distribution and we know the generating function (or the moment generating function), that is, we know

$$G_X(\alpha) = E[e^{-\alpha X}] = e^{-\lambda(1-\exp(-\alpha))}.$$

Taking derivatives at $\alpha = 0$ we obtain

Table 2.8: R code for the method of maximum likelihood.

```
# the original data is grouped, use the following command to ungroup them:
xpoi=NULL
for (i in min(x[,1]):max(x[,1]))
        xpoi=c(xpoi,rep(i,x[i+1,2]))
xpoi

# now, to apply the maximum likelihood method.

#Option 1
library(stats4) # Loads stats4.
x=NULL
#likelihood function
ll=function(lambda) {
                n=length(xpoi)
                x=xpoi
                -(-lambda*n+sum(x)*log(lambda))
                }
est1=mle(minuslog=ll, start=list(lambda=2))
summary(est1)

#Option 2 (simplest option)
library(MASS)
fitdistr(xpoi,"poisson")
```

$$\mu_1 = E[X] = -\left.\frac{dG_X}{d\alpha}\right|_{\alpha=0} = \lambda.$$

This identity asserts that in order to determine the parameter of the distribution from the sample data, all we need is to estimate the first moment (the expected value) of the random variable, that is

$$\widehat{\lambda} = \frac{1}{n}\sum_{k=1}^{n} x_k.$$

We are happy because the same result is obtained as when applying the maximum likelihood method.

2.3.2 Binomial distribution

Suppose that we have data describing the number of daily failures at some power generating plant. The data listed in Table 2.9 was simulated from a binomial $B(0.2, 60)$. The power generation failures may be the of cause of failures in a network of banks, or the failure of some other process depending on the availability of electrical power.

2.3 Examples — 19

Table 2.9: Number of failures at a power plant.

Number of events (failures) k	Frequencies (n_k)	Relative frequencies (p_k)
6	8	0.08
7	19	0.19
8	36	0.036
9	60	0.060
10	86	0.087
11	110	0.111
12	125	0.126
13	128	0.129
14	118	0.119
15	99	0.100
16	76	0.077
17	53	0.053
18	35	0.053
19	21	0.021
20	11	0.011
Total	985	–

In Figure 2.4 we see the bar diagram corresponding to the data in Table 2.9.

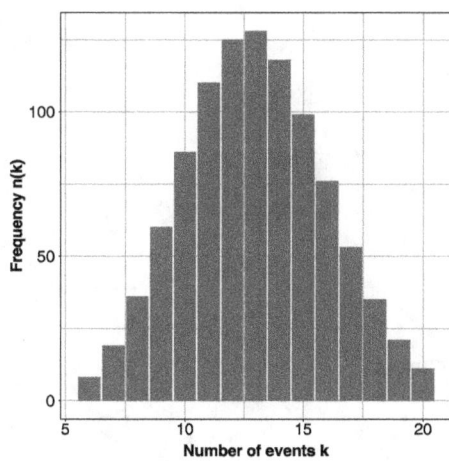

Figure 2.4: Bar graph of the daily failures at a power plant.

To apply the Panjer procedure we created Table 2.10 with a new column corresponding to the values of $k \frac{\hat{p}_k}{\hat{p}_{k-1}} = k \frac{n_k}{n_{k-1}}$.

In Figure 2.5 we display the plot of k versus the values of $k \frac{\hat{p}_k}{\hat{p}_{k-1}}$, or, equivalently, $k \frac{n_k}{n_{k-1}}$. Clearly, the points follow a straight line with negative slope, which suggests that the data comes from a binomial distribution and we only need to determine its parameters.

Table 2.10: Number of failures at a power plant.

Number (k) of events	Frequency (n_k)	$k\frac{n_k}{n_{k-1}}$
6	8	–
7	19	16.62
8	36	15.15
9	60	15
10	86	14.33
11	110	14.06
12	125	13.63
13	128	13.31
14	118	12.90
15	99	12.58
16	76	12.28
17	53	11.85
18	35	11.88
19	21	11.40
20	11	10.47
Total	985	–

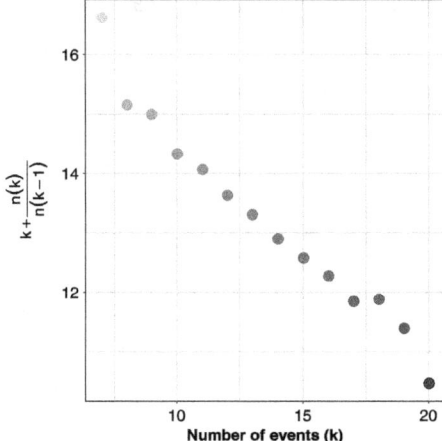

Figure 2.5: Plot of k versus $k\frac{n_k}{n_{k-1}}$.

We fit the data to the line $k\frac{p_k}{p_{k-1}} = k\frac{n_k}{n_{k-1}} = ka + b$ and obtain $a = -0.3937$ and $b = 18.5668$, with the resulting line plotted in Figure 2.6.

We already saw that the relationship between the parameters a, b and the parameters p, m is given by

$$a = -\frac{p}{1-p}, \quad b = (m+1)\left(\frac{p}{1-p}\right), \quad \text{and} \quad p_0 = (1-p)^m.$$

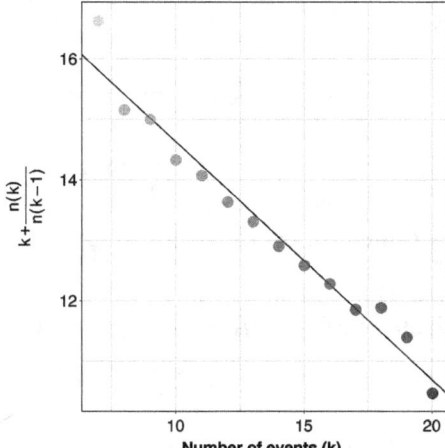

Figure 2.6: Linear regression.

Solving this simple system, we obtain the following parameter values: $p = 0.282248855$, $n = 46.15977$ and $p_0 = 2.214 \times 10^{-7}$. We can supplement this analysis by invoking the method of maximum likelihood to provide another determination of the parameters, and by invoking the χ^2-test to complete the analysis.

The likelihood function of a dataset following the binomial distribution $P(X = x) = \binom{m}{x} p^x q^{m-x}$ is

$$L(\theta; x) = \prod_{i=1}^{n} \binom{m}{x_i} p^{x_i} q^{m-x_i} \simeq \prod_{i=1}^{n} p^{x_i} q^{m-x_i} = p^{\sum_{i=1}^{n} x_i} (1-p)^{\sum_{i=1}^{n}(m-x_i)}.$$

We dropped the term not containing p and wrote ($\simeq$) to represent that. The log-likelihood is

$$\log L(\theta; x) = \sum_{i=1}^{n} x_i \log(p) + \sum_{i=1}^{n} (m - x_i) \log(1 - p).$$

Differentiating with respect to p and equating to 0 we obtain

$$\frac{\sum_{i=1}^{n} x_i}{p} - \frac{\sum_{i=1}^{n} (m - x_i)}{1 - p} = 0,$$

or,

$$p = \frac{\sum_{i=1}^{n} x_i}{nm} = \frac{\bar{X}}{m}.$$

Therefore, $\hat{p} = 0.2$, which coincides with the values determined previously. We leave the application of the χ^2-test as homework for the reader.

3 Individual severity models

In this chapter we shall examine a few of the large variety of possible continuous, positive random variables that are used to model individual loss severities. Besides the obvious listing of their densities and statistical properties, of potential interest for us is the fact that the Laplace transform of the probability density (in short, the Laplace transform of the random variable) can be explicitly computed. Let us begin with the basic definition.

Definition 3.1. Let X be a positive random variable with density $f_X(x)$. Its Laplace transform $\psi(\alpha)$ is defined for any $\alpha > 0$ by

$$\phi(\alpha) = E[e^{-\alpha X}] = \int_0^\infty e^{-\alpha x} f_X(x) dx.$$

A widely used and related concept is the moment generating function, given by $M(t) = \phi(-t)$. Those who get annoyed by the minus sign when computing derivatives prefer this notation.

Even though sometimes the range of the definition can be extended to a larger subset of the real line, we shall not bother about this too much. Further properties of the Laplace transform will be explored in Chapter 6. The definition given is all we need for the time being. In all cases we will make explicit the relationship between the moments of the variable and the parameters of the distribution.

3.1 A short catalog of distributions

The densities considered below are characterized by parameters that are easily related to moments of the variables, making the problem of their estimation easy. The reader is supposed to finish the incomplete computations.

3.1.1 Exponential distribution

The exponential distribution is characterized by a single parameter λ. The density is given by

$$f(x) = \lambda e^{-\lambda x}, \qquad M(t) = \lambda(\lambda - t)^{-1},$$
$$F(x) = 1 - e^{-\lambda x}, \qquad \phi(\alpha) = \frac{\lambda}{\lambda + \alpha}. \tag{3.1}$$
$$E[X^k] = \frac{k!}{\lambda^k}; \quad k \in \mathbb{N},$$

The reader should verify that for any pair of positive numbers t, s the following holds: $P(X > t + s \mid X > t) = P(X > s)$. This property actually characterizes the exponential density.

3.1.2 The simple Pareto distribution

This density is characterized by two positive parameters λ and θ (which is usually supposed to be known because it denotes a cutoff point),

$$f(x) = \lambda \theta^\lambda x^{-(\lambda+1)}; \quad x > \theta,$$
$$F(x) = 1 - (\theta/x)^\lambda,$$
$$E[X^k] = \frac{\lambda \theta^k}{(\lambda - k)}; \quad k \in \mathbb{N}, \, k < \lambda, \quad (3.2)$$
$$\phi(\alpha) = \lambda(\theta\alpha)^\lambda \Gamma(-\lambda, \alpha).$$

This is one of the simplest examples in which the Laplace cannot be given by a close analytic expression. The series expansion (in α) of the incomplete gamma functions is nevertheless convergent. This makes the need for the approach that we favor apparent, as the computations to obtain aggregate losses involving individual severities of the Pareto type have to be carried out numerically. Note as well that when θ is known, λ can be obtained by estimating the mean.

3.1.3 Gamma distribution

The density of such variables is characterized by $\lambda > 0$ and $\beta > 0$, and its simple properties are:

$$f(x) = \frac{(x\beta)^\lambda e^{-x\beta}}{x\Gamma(\lambda)},$$
$$E[X^k] = \frac{\beta^{-k}\Gamma(\lambda + k)}{\Gamma(\lambda)}; \quad k > -\lambda,$$
$$E[X^k] = \beta^{-k} \prod_{j=1}^{k}(\lambda + k - 1); \quad \text{when } k \in \mathbb{N},$$
$$\phi(\alpha) = \left(\frac{\beta}{\beta + \alpha}\right)^\lambda; \quad \alpha > 0.$$

3.1.4 The lognormal distribution

This is a very widely used model. It is also interesting for us since its Laplace transform has to be dealt with numerically. When X is lognormal, the $\ln(X)$ is distributed according to a normal distribution, hence its name. The basic properties of the variable are:

$$f(x) = \frac{1}{x\sigma\sqrt{2\pi}}\exp\left(-\frac{1}{2}\left(\frac{\ln x - \mu}{\sigma}\right)^2\right) = \frac{\phi(\frac{\ln x - \mu}{\sigma})}{x\sigma},$$

$$F(x) = \Phi\left(\frac{\ln - \mu}{\sigma}\right), \quad (3.3)$$

$$E[X^k] = \exp\left[k\mu + \frac{1}{2}k^2\sigma^2\right].$$

The Laplace transform of this density is a complicated affair. Try summing the series expansion of

$$E[e^{-\alpha X}] = \sum_{k=0}^{\infty} \frac{(-\alpha)^k}{k!} E[X^k],$$

where the moments are given a few lines above. Again, it may be easier to proceed numerically: Simulate a long sequence of values of X and compute the integral by invoking the law of large numbers.

3.1.5 The beta distribution

This is a standard example of a continuous distribution on a bounded interval. It is described by two shape parameters a, b, and a parameter L that specifies its range, which we chose to preserve as generic, for when it is unknown it can be randomized. Otherwise, it can be scaled to $L = 1$. The properties of the beta density are:

$$f(x) = \frac{\Gamma(a+b)}{\Gamma(a)\Gamma(b)}\frac{x^{a-1}(L-x)^{b-1}}{L^{a+b-1}}; \quad 0 < x < L,$$

$$E[X^k] = L^k \prod_{j=1}^{k}\frac{(a+k-j)}{(a+b+k-j)}; \quad k \in \mathbb{N},$$

$$E[X^k] = \frac{L^k \Gamma(a+b)\Gamma(a+k)}{\Gamma(a)\Gamma(a+b+k)}; \quad k > -a, \quad (3.4)$$

$$\phi(\alpha) = \sum_{k=1}^{\infty}\left(\prod_{j=0}^{k-1}\frac{a+j}{a+b+j}\right)\frac{(-\alpha L)^k}{k!}.$$

Simple as this density may appear, observe that its Laplace transform is a again given by a power series.

3.1.6 A mixture of distributions

We can think about mixtures as we think of the conditional expectations of a random variable given some other variables. This certainly is an extension of the notion of Poisson mixture considered in Chapter 2. A motivation for this topic comes from the fact that in many instances, when observing losses, we may not distinguish the source or the cause of the loss. For example, when totaling collision damage payments, an insurance company has the record of all claims paid, but not necessarily discriminated by gender or age group.

Here we shall consider only the case of discrete mixtures. To establish the necessary notation, consider a discrete random variable Y taking values $\{y_1, \ldots, y_k\}$ with probabilities $P(Y = y_i) = w_i$. Consider as well random variables X, X_1, X_k such that

$$P(X \leq x \mid Y = y_i) = F_{X_i}(x); \quad i = 1, \ldots, k.$$

With all these notations, we define the finite mixture as follows:

Definition 3.2. We say that the random variable X is a finite mixture of the random variables $X_1, \ldots, X_k$ with weights w_i whenever

$$F_X(x) = P(X \leq x) = \sum_{i=1}^{K} w_i F_{X_i}(x). \tag{3.5}$$

An application

A simple but interesting application of this concept to the analysis of loss distributions goes as follows. The losses in some line of business occur in two separate categories, or two event types: low severity or of high severity. When the severity is low, the mean loss is small, and when the severity is high, the mean is 'very' large. Suppose for example that the severity distribution in both cases is given by a gamma type density, and that the frequencies of occurrence of each type of event are 99 % and 1 % respectively. We can then write

$$F_X(x) = 0.99\Gamma(a, b)(x) + 0.01\Gamma(500a, b)(x),$$

where we supposed that the mean in the high severity case is 500 times the mean of the low severity case, and that the shape parameter b is the same.

3.2 Numerical examples

In this section we consider two simple numerical examples, representative of the type of analysis carried our in order to determine the distribution of individual losses to establish a model of aggregate losses.

3.2.1 Data from a density

Consider the data gathered in Table 3.1, in which we list the events and the loss occurring in each event. We want to determine the statistical distribution of such losses.

Table 3.1: Observed severities.

Event (k)	Loss	Event (k)	Loss	Event (k)	Loss
1	1 516 028.32	26	453 455.18	51	221 558.05
2	1 498 141.46	27	449 911.93	52	203 935.81
3	1 436 992.93	28	446 218.58	53	191 718.66
4	1 411 135.84	29	399 945.39	54	185 600.43
5	1 402 875.75	30	396 987.916	55	152 676.24
6	1 382 964.54	31	390 013.44	56	147 058.98
7	1 243 413.57	32	389 501.23	57	145 930.81
8	1 170 903.11	33	378 440.99	58	143 431.94
9	976 245.85	34	369 399.32	59	134 381.89
10	969 018.21	35	364 209.60	60	125 367.85
11	871 878.61	36	361 308.27	61	118 109.20
12	755 416.05	37	337 976.11	62	115 076.13
13	730 944.58	38	327 905.81	63	112 081.67
14	669 860.28	39	323 704.58	64	93 575.70
15	668 858.27	40	319 140.11	65	92 119.22
16	641 890.75	41	316 762.41	66	86 061.66
17	621 547.09	42	304 855.13	67	83 937.03
18	599 481.55	43	296 318.58	68	55 383.49
19	582 070.72	44	272 157.26	69	49 195.29
20	547 046.03	45	270 419.08	70	48 974.33
21	524 929.89	46	251 966.24	71	42 990.75
22	494 375.38	47	244 035.60	72	25 507.72
23	482 097.24	48	232 145.86	73	11 115.65
24	458 880.05	49	223 349.34	74	455.52
25	456 639.17	50	222 457.75		

An exploratory analysis of the data is the first step in determining the frequency distribution behind the data. The standard procedure consists of computing the simple statistics and analyzing the visual displays of the data to get an idea of the type of density. This visual display also suggests possible densities to compare against the data. After that, a goodness of fit technique, like the Kolmogorov–Smirnov test, can be applied to measure the quality of the fit.

In Figure 3.1 we show the histogram of the data in Table 3.1. The obvious step is to compare it with the plot of a few densities that usually appear in similar situations. Since the histogram suggests decreasing frequency as severity increases, we consider distributions like the exponential or the gamma to begin with.

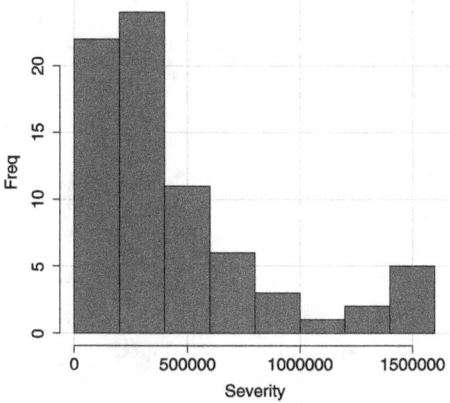

Figure 3.1: Severity histogram.

In Figure 3.2 we compare the distribution of the data to the exponential density of parameter $\lambda = \frac{1}{\bar{X}}$ and to the gamma distribution of parameters $\alpha = \frac{\bar{X}^2}{S^2}$ and $\beta = \frac{\bar{X}}{S^2}$, where $\bar{X}$ and S^2 are, respectively, the sample mean and the sample variance of the data. Recall that the relationship between the parameters and the statistics for these distributions is given by $E[X] = \frac{1}{\lambda}$ for the exponential density, whereas $E[X] = \frac{\alpha}{\beta}$ and $\text{Var}[X] = \frac{\alpha}{\beta^2}$ for the gamma density.

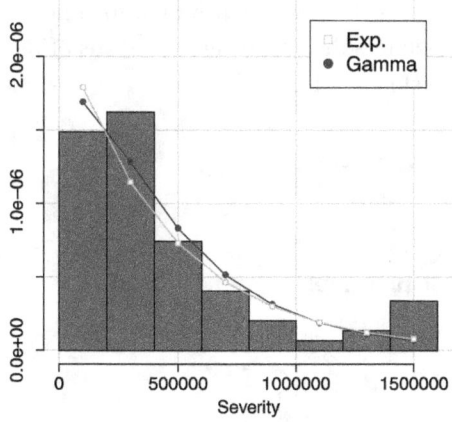

Figure 3.2: Comparison with theoretical densities.

We can also compare the cumulative distribution functions to the cumulative empirical distribution. This is presented in Figure 3.3, again for the same two sets of parameters mentioned above.

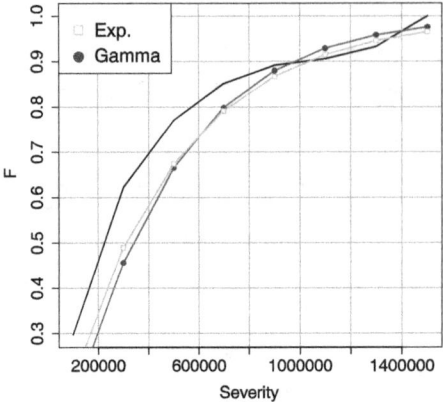

Figure 3.3: Comparison of cumulative distribution functions.

Another way of comparing data against theoretical distributions is by means of a quantile plot (Q-Q plot), which is a graphical procedure to diagnose differences between a probability distribution of a sample and a possible distribution that the sample may come from. If the sample comes from the trial theoretical distribution, the plot should be 'almost' linear.

In Figure 3.4 we display the (Q-Q) plot of the data and the gamma and exponential densities. A line at 45° was drawn to facilitate the comparison. Certainly, to suppose that the data comes from an exponential distribution makes more sense than to suppose that it comes from a gamma distribution. To estimate the parameter by means of the maximum likelihood procedure, we form the likelihood function

$$L(\theta; x) = \prod_{i=1}^{n} \lambda e^{-\lambda x_i} = \lambda^n e^{-\lambda \sum_{i=1}^{n} x_i}$$

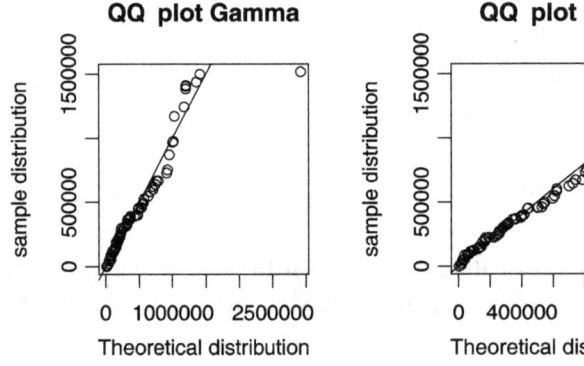

Figure 3.4: Q-Q plots.

corresponding to an exponential density $f(x) = \lambda e^{-\lambda x}$. Therefore, the log-likelihood is given by

$$\log L(\theta; x) = n \log(\lambda) - \lambda \sum_{i=1}^{n} x_i.$$

Equating the derivative with respect to λ to 0, we obtain

$$\lambda = \frac{n}{\sum_{i=1}^{n} x_i}.$$

After having estimated the parameter, we have to make sure that the data is really a sample from that distribution. For that, the standard procedure consists of using goodness of fit tests, like the χ^2 or the Kolmogorov–Smirnov tests. Once more, we direct the reader to Chapter 14 for a more detailed description of the statistical procedures that we use.

When the data $\{x_1, \ldots, x_n\}$ is supposed to come from an exponentially distributed random variable, we perform the following hypothesis test: We compare the null hypothesis $H_0: X \sim \text{Exp}(\lambda)$ against the alternative $H_1: X$ does not follow a $\text{Exp}(\lambda)$.

To apply the test, we carry out the steps described in Table 3.2. That is, we place the data in increasing order and compute the empirical distribution (F_n) at consecutive points, as well as the theoretical distribution F_0, and then the differences indicated in the first row of the table. We then apply the Kolmogorov–Smirnov test as described in Chapter 14. The value of the test statistic is $D = 0.080342$, which is to be compared to the critical value $d_{0.05,75} = 0.15$. Since $D < d_{0.05,75}$ we do not reject the null hypothesis, that is we stick to the exponential distribution as suggested by the (Q-Q) plot.

Table 3.2: Data for the Kolmogorov–Smirnov test.

i	x_i	$F_n(x_i)$	$F_n(x_{i-1})$	$F_o(x_i)$	$|F_n(x_i) - F_o(x_i)|$	$|F_o(x_i) - F_n(x_{i-1})|$
1	455.52	0.01351351	0.00000000	0.001019638	0.0124938758	0.0010196377
2	11 115.65	0.02702703	0.01351351	0.024586672	0.0024403552	0.0110731583
3	25 507.72	0.04054054	0.02702703	0.055524583	0.0149840426	0.0284975561
4	42 990.75	0.05405405	0.04054054	0.091789960	0.0377359056	0.0512494191
5	48 974.33	0.06756757	0.06756757	0.103879227	0.0363116590	0.0498251726
⋮	⋮	⋮	⋮	⋮	⋮	⋮
70	1 402 875.75	0.94594595	0.93243243	0.956795027	0.0108490807	0.0243625943
71	1 411 135.84	0.95945946	0.94594595	0.957586920	0.0018725390	0.0116409745
72	1 436 992.93	0.97297297	0.95945946	0.959973219	0.0129997538	0.0005137597
73	1 498 141.46	0.98648649	0.97297297	0.965095910	0.0213905767	0.0078770632
74	1 516 028.32	1.00000000	0.98648649	0.966466477	0.0335335235	0.0200200100

4 Some detailed examples

The contents of this chapter are a commented expansion of the list of examples mentioned right after (1.1) in Chapter 1. Even though the statements are variations on a common theme, it is nevertheless nice to see how seemingly different questions lead to similar mathematical problems.

4.1 Claim distribution and operational risk losses

Even though the Basel II requirements for the calculation of regulatory capital were an important motivation for the calculation of the distribution of operational risk losses, the change in the new (as of 2016, see [71]) requirements has changed the motivation but not the need to tackle the problem. This is first because the problem is interesting in itself; second because it is also an important problem in the insurance industry; and third because the same problems appears when studying credit losses. The calculation of the density of the distribution of claims is the first step in the computation of risk premia, and the calculation of the distribution of credit losses is the first step in the computation of regulatory capitals and the determination of lending rates.

The nice thing about the Basel II requirements for the calculation of regulatory capital for operational risk is that the Basel Committee provided banks with a categorization of risk as a detailed list of risk types per business line. Not only that, each risk type and business line had a fine structure in which many of the risk events to measure (and pay attention to) were specified. This made life easier for both the risk manager at the bank and the regulator.

For the risk analyst life was 'simpler' too, because they had a clear order for risk aggregation and model building: for as long a length of time as possible, observe the events in each category; record the corresponding loss; aggregate first by business line, and then by risk type; and then aggregate once more to obtain the total loss. After this, the job becomes to determine the distribution of the aggregate losses, and from that, compute the risk capital.

Even though regulatory capital may now be calculated according to a different methodology, the categorization provided by the Basel Committee may be useful for risk management, for it tells the risk manager where and what to look at, for risk prevention purposes.

To come back to (1.1), let $N_1, \ldots, N_H$ be the integer valued random variables describing the different possible types of risk events during the observation period. For example, if we take the number of fraudulent events with credit cards at ATM machines in a given town, then h labels the ATM machines. Or $\{N_h, h = 1, \ldots, H\}$ may be the number of claim reports due to car crashes classified by age group or by gender, or the number of bad checks paid at a certain branch of a bank.

Certainly, the total number of events during the period is $N = \sum_{h=1}^{H} N_h$. The issue that the analyst has to attend to here is how to model dependence among the different N_h. In many situations independence is the natural assumption, but that is not so in all situations.

Next comes the choice of the model for the individual losses. As we said, for many applications it is convenient to use continuous valued random variables. So, let $\{X_{h,k} \mid h = 1,\ldots,H; k \geq 1\}$ be a collection of random variables modeling the k-th loss (amount of claim) of the h-th type. It is also reasonable to suppose that the number of losses of each type is unbounded. Of course, anybody can see that the number of car collisions cannot be larger than some function of the number of inhabitants of a country, or that the number of credit card frauds cannot be larger than some multiple of the number of card holders, each of which is large but finite. But for modeling purposes it is easier to regard these as unbounded. With all this, the total loss is modeled by

$$S = \sum_{h=1}^{H} \left(\sum_{k=1}^{N_h} X_{h,k} \right). \tag{4.1}$$

We added round brackets to (4.1) just to emphasize the two levels of aggregation implied within, which introduces the problem of dependency among levels. Again, the losses within each type are usually modeled by independent, identically distributed random variables, and independent of the number of events, such independence may or may not be assumed when aggregating different risk types, or when aggregating severities across business lines. A similar comment applies in the insurance industry. For example, when aggregating damage caused by natural disasters, it is clear that there must be some sort of regional aggregation, even if losses by similar events in geographically separated areas may most of the time be regarded as independent.

Of course, it is clear that if we do not separate among the different risk types, and have one single random variable N to count all risk events during the time period, and just a standard family $\{X_k : k \geq 1\}$ of individual severities, then the compound loss in (4.1) looks like (1.1), that is,

$$S = \sum_{k \geq 1} X_k.$$

In Chapter 1 we also commented on a converse problem, and now we have introduced notation to make the comment explicit. It may be important and interesting that given $N = \sum_h N_h$, and with S given by (4.1), to be able to say as much as possible about the N_h and the $X_{h,n}$ themselves. This theme will be taken up again in Chapter 12.

4.2 Simple model of credit risk

We begin by clarifying the meaning of the title. Credit risk models any kind of risk occurring when one of the parties to any contract defaults on their (contractual) obligations, causing some sort of monetary loss to the other. The simplest example of this situation occurs when of one or more of the borrowers at a bank do not return back the amount of money agreed upon in the contract.

The simplest situation is the following. Suppose that a given bank grants at the beginning of the year a number of identical loans to n creditors, each of which has to return the amount $(1 + r_l)L$ by the end of the year. We denote by r_l the rate at which the bank is allowed to lend money, and by r_b the rate at which the central bank borrows money (at the so called risk free rate). Suppose, furthermore, that all customers have the same probability $p > 0$ of not returning the money. If a customer does not pay their loan, then the bank looses the amount lent (plus the interest rate) that it was going to collect. To examine the alternatives to lending money at risk, we note that instead of granting individual loans the bank could have lent the money to the central bank, receiving $n(1+r_b)L$ with certainty. Hence the name risk free rate for r_b.

We will denote by D_i, $i = 1, \ldots, n$ a binary random variable such that $D_i = 1$ means the i-th customer defaults, and $D_i = 0$ means the customer pays their debt. The loss to the bank (given that) the i-th customer defaults is $(1 + r_l)L$, which written as a random variable is $(1 + r_l)D_i L$. This means the total (random) loss (or money not collected at the end of the year) is given by $\sum_i D_i(1 + r_l)L$, and the distribution of this variable is easy to obtain because it is essentially distributed according to a binomial law.

At this point we come across a couple of interesting issues in risk management. First: Is there a fair way to choose r_l? One possible interesting way to answer the question is that the bank chooses r_l in such a way that the money it expects to receive in each of the two possible investment alternatives is the same. As in this simple model $E[\sum_i D_i] = np$, we have

$$E\left[\sum_{i=1}^n (1 - D_i)(1 + r_l)L\right] = n(1 - p)(1 + r_l)L = n(1 + r_b)L.$$

That is, the bank adjusts the lending rate to a level at which the expected amount to be received from lending at risk to individual borrowers equals the amount received by lending at zero risk (at the rate r_b). A simple calculation shows that the spread (the difference in rates) is

$$r_l - r_b = \frac{p}{1 - p}(1 + r_b)$$

But this is not a good managerial decision, because the number of defaults may fluctuate about its mean (np) in a way that makes large losses highly probable. A criterion for designing a better way to determine the lending rate r_l goes as follows: Fix a large fraction $0 < f < 1$ of sure earnings and a large probability $0 < \alpha < 1$ with which you want your earnings to overtake $fn(1 + r_b)L$ and choose r_l such that this is true. That is, choose r_l so that

$$P\left(\sum_{i=1}^{n}(1-D_i)(1+r_l)L > fn(1+r_b)L\right) \geq \alpha.$$

As $\sum_{i=1}^{n}(1-D_i)$ is a binomial distribution with parameters n, $(1-p)$, it is easy to see that such an r_l can be found. Even though the example is too simple, we already see that the need to compute the distribution of losses (or defaults) is an important part of the analysis.

To make the model a bit more realistic, suppose that the bank has several categories of creditors, for example large corporations, middle size corporations, small businesses and individual borrowers. Suppose as well that the creditors within each category have the same probability of default. It is reasonable to suppose that the number of large corporations and middle size corporations is not very large, and it is also reasonable to suppose that the random number among them that default may be described by a binomial model. For small businesses and individual borrowers, if the individual probabilities of default are small, it may make sense to model the number of defaults by variables of the Poisson type.

The total loss to a bank caused by defaulting creditors may be described by

$$S = \sum_{k=1}^{M}\sum_{n=1}^{N_k} X_{k,n}, \quad (4.2)$$

where N_k is the random number of defaults of the creditors in the k-th category, and $X_{k,n}$ is the loss to the bank produced by the n-th borrower in the k-th category. As mentioned, each of the M summands has some internal structure and the necessary modeling involves quite a bit of analysis. For example, if we consider large corporations, there is information about their credit ratings from which an individual probability of default may be extracted. As the corporations function in a common economic environment, the probabilities of default may not be independent. Other sources of credit losses to aggregate may come from loans to small corporations or small businesses, the residential mortgage loans, the standard consumer loans via credit card, and so on. Clearly each group has internal similarities and possible dependencies.

Besides that, the notion of the occurrence of default depends on the type of borrower. To model the default by a corporation, or by a collection of them in a credit portfolio, a variety of ways have been proposed to model the occurrence of default, and then to model the probability of default of a collection of corporations. Such details have to be taken into account when trying to determine the distribution of credit losses affecting a bank. For details about such modeling process, the reader may consult chapters 8 and 10 in [69].

We mention as well that the model in (4.2) is the starting point for a methodology (called *Credit Risk+*, CR+) proposed by the Credit Suisse Bank, which extends the simple model presented at the beginning of this section into a slightly more elaborate model, and still leads to an analytic computation of the Laplace transform of the total loss.

The extension involves several changes. Instead of allowing all obligors to have (or to cause) independent and identically distributed (IID) individual losses, it maintains a given fixed number of obligors, which may have independent but non identically distributed losses. The variation on the theme described above proposes a random number of defaults, which are grouped into classes such that the individual losses within each class are IID. To continue with the CR+ model proposal we write

$$S = \sum_{n=1}^{N} D_n E_n,$$

where D_n is the default indicator introduced above and $E_n = \text{LGD}_n \times \text{EAD}_n$ is the loss caused by the n-th obligor, where LGD_n stands for 'loss given default' and EAD_n stands for 'exposure at default'. It is decomposed into two factors: an *exposure at default*, which quantifies the maximum loss, and the *loss given default*, which is sometimes written as $1-f_n$, where f_n stands for the fraction recovered after the default of the n-th obligor. But let us stick to the simpler notation and let E_n denote the random variable modeling the loss caused by the default of the n-th obligor. Usually a base unit E of loss is introduced and we set $v_n = E_n/E$ for the number of units of loss caused by the n-th obligor. Denoting S/E by S once more we have

$$S = \sum_{n=1}^{N} D_n v_n = \sum_{n=1}^{N} \lambda_n,$$

where λ_n denotes the loss (in units of E) produced by the n-th obligor. The second part of the CR+ proposal consists of supposing that the default frequencies are Poisson instead of Bernoulli. That is justified by the proponents of the model as follows: If $P(D_n = 1) = p_n$ is very small, then

$$\psi_{\lambda_n}(\alpha) = E[e^{-\alpha \lambda_n}] = (1 - p_n + p_n E[e^{-\alpha v_n}])$$

$$e^{\ln(1-p_n+p_n E[e^{-\alpha v_n}])} \approx e^{-p_n(1-E[\exp(-\alpha v_n)])}.$$

That is, the default indicators now are now supposed to be integer valued random variables distributed according to a Poisson(p_n). Since in a real economy, the defaults may in some way be affected by a common factor, in order to introduce dependence and still keep a simple analytical model the last extension to the simple model consists of replacing p_n by the following simple possible mixture model:

$$p_n \left(\sum_{k=0}^{K} w_{nk} X_k, \right) \quad \text{with } w_{nk} \geq 0, \quad \text{and} \quad \sum_{k=1}^{K} w_{nk} = 1,$$

where the random factors $X_k \sim \Gamma(\xi_k, 1/\xi_k)$, and $X_0 = 1$. This means the random factor is positive and has a mean equal to 1. But, to add to the unpalatable nature of the model, as it admits possible multiple defaults by the same obligor, there is also the possibility of the default probability to being larger than 1. To finish, let us introduce the notation $P_k(\alpha) = \sum_{n=1}^{N} p_n w_{nk} \mathrm{E}[e^{-\alpha v_n}]$. With this, the Laplace transform of the total loss can be easily seen to be given by

$$\mathrm{E}[e^{-\alpha S}] = \prod_{k=1}^{K} \left(\frac{1}{1 + \xi_k P_k(\alpha)} \right)^{\xi_k}.$$

Again, the problem is now to invert this Laplace transform.

4.3 Shock and damage models

These are restricted versions of a model that is inherently dynamical. The accumulation of damage produced by shocks is a stochastic process, which at any given time is described by a compound random variable like those described above. Considering only the accumulated damage in a given fixed period of time is a very small aspect of the problem. The literature on this type of problems is large. As an introduction to the field, consider [75].

Consider N to be the number of events, or shocks, that affect a structure during a fixed period of time. At the time that the n-th shock occurs an amount of wear (or damage) W_n is accumulated. It is natural to suppose that the W_n are independent and identically distributed. Actually, and like in the case of credit risk, there is some internal structure to W_n. The resulting damage occurs when the intensity of the shock is larger than some threshold, and we may think of W_n as the response of the system to the intensity of the shock. Anyway, the 'severity' of the accumulated damage is again given by a compound variable

$$S = \sum_{n=0}^{N} W_n.$$

Even though the question is more natural in a dynamical setup, the reason that the distribution of S is relevant is because as soon as the accumulated damage becomes larger than some critical value K, the effect of the damage entails serious losses of some kind, and this involves the computation of the distribution of accumulated damage.

4.4 Barrier crossing times

Let us examine a few dynamical versions of the problems described above. Let us denote by $N(t)$ an integer valued process, usually supposed to have independent, identically distributed increments, and let $\{X_n \mid n \geq 1\}$ be a collection of independent identically distributed positive random variables. Consider either of the following processes:

$$S(t) = s_0 + \sum_{n=1}^{N(t)} X_n - rt$$

or

$$S(t) = s_0 + rt - \sum_{n=1}^{N(t)} X_n.$$

When $r = 0$, the first example can be used to describe accumulated damage in time. When $r > 0$, we can think of $S(t)$ as the savings in an account, or the water level in a dam, in which there is an 'input' of money or rain at random times described by $N(t)$, and r quantifies the spending or release rates. The constant (or perhaps random) term s_0 models the initial amount of whatever it is that $S(t)$ denotes. The second example is used in the insurance industry to denote the capital of the insurance company. Naturally, s_0 is the initial capital, r the rate at which premia are collected and the compound variable is the amount paid in claims.

Apart from the simplifications present in each of the models, for us the interest is in a question that appears in all of them: When is the first time some critical barrier is reached? This is defined by

$$T = \inf\{t \geq 0 \mid S(t) \geq K\}$$

or perhaps

$$T = \inf\{t \geq 0 \mid S(t) \leq K\}$$

for some level K related to the nature of the problem. The problem of interest in many applications is to determine the distribution (usually a probability density) of T. The relationship of this problem with the problems described in the three sections above is that the methodology to solve it is similar to the methodology of determining the density of the compound variables. In many cases, the nature of the inputs in the problem is such that a differential or integral equation to compute

$$\phi(\alpha) = E[e^{-\alpha T}]$$

can be established and solved.

An important class of such problems appears when dealing with the distribution of defaults of corporations to compute credit losses. A class of models, called structural models, proposes that defaults occur when the value of a corporation falls below a certain threshold related to the liabilities of the corporation. When dealing with a portfolio of corporations we have a many-dimensional barrier crossing problem.

4.5 Applications in reliability theory

Another branch of applied mathematics in which the problem of determining the distribution of a positive random variable crops up in a natural way is reliability theory. The failure rate of a complex system is a function of the failure rates of its components. The components are chained in groups of series and/or parallel blocks, which may or may not fail independently.

If a module consists of K-blocks placed in series, and their times to failure are random variables T_i (not necessarily identically distributed), the time to failure of the block is given by $\min\{T_i: i = 1, \ldots, K\}$. Similarly, if a component is made up of k-blocks connected in parallel, and their times to failure are denoted by T_i, and if the system does not fail until all subcomponents have stopped working, then the time to failure of the component is given by $\max\{T_i: i = 1, \ldots, k\}$.

It is then clear that if $T_1^c, \ldots, T_M^c$ are the failure times of the subcomponents of a system, the failure time of the system is some more or less complicated function $H(T_1^c, \ldots, T_M^c)$ of the failure times of the components. Note that even when we know the distributions of the blocks, and are lucky enough that the distribution of failure time of the subsystems can be computed, our luck does not usually extend to the computation of the distribution of the failure time of the system. For that we can now resort to the Monte Carlo simulation of a large sample of $H(T_1^c, \ldots, T_M^c)$ and try to obtain the distribution that fits the resulting histogram, or to use the Laplace transform based techniques that we propose in this volume. Of course, we may have empirical data and we may start from that, without having to resort to simulations.

5 Some traditional approaches to the aggregation problem

5.1 General remarks

In Chapters 2 and 3 we considered some standard models for the two basic building blocks for modeling the compound random variables used to model risk. First, we considered several instances of integer valued random variables used to model the frequency of events, then some of the positive continuous random variables used to model individual losses or damage each time a risk event occurs.

From these we obtain the first level of risk aggregation, which consists of defining the compound random variables that model loss. Higher levels of loss aggregation are then obtained by summing the simple compound losses.

As we mentioned in the previous chapter, the first stage of aggregation has been the subject of attention in many areas of activity. In this chapter we shall review some of the techniques that have been used to solve the problem of determining the distribution of losses at the first level of aggregation. This is an essential step prior to the calculation of risk capital, risk premia, or breakdown thresholds of any type.

The techniques used to solve the problem of determining the loss distribution can be grouped into three main categories, which we now briefly summarize and develop further below.

- Analytical techniques: There are a few instances in which explicit expressions for the frequency of losses and for the probability density of individual losses are known. When this happens, there are various possible ways to obtain the distribution of the compound losses. In some cases this is by direct computation of the probability distribution, in other cases by judicious use of integral (mostly Laplace) transforms. Even in these cases, in which the Laplace transform of the aggregate loss can be obtained in close form from that of the frequency of events and that of the individual severities, the final stage (namely the inversion of the transform) may have to be completed numerically.
- Approximate calculations: There are several alternatives that have been used. When the models of the blocks of the compound severities are known, approximate calculations consisting of numerical convolution and series summation can be tried. Sometimes some basic statistics, like the mean, variance, skewness, kurtosis or tail behavior or so are known, in this case these parameters are used as starting points for numerical approximations to the aggregate density.
- Numerical techniques: When we only have numerical data, or perhaps simulated data, obtained from a reliable model, we can only resort to numerical procedures.

5.1.1 General issues

As we proceed, we shall see how the three ways to look at the problem overlap each other. Under the independence assumptions made about the blocks of the model of total severity given by the compound variable $S = \sum_{n=1}^{N} X_n$, it is clear that

$$P(S \leq x) = \sum_{k=0}^{\infty} P(S \leq x \mid N = k) P(N = k)$$

$$= \sum_{k=0}^{\infty} P\left(\sum_{n=1}^{k} X_n \leq x \mid N = k\right) P(N = k)$$

$$= \sum_{k=0}^{\infty} P\left(\sum_{n=1}^{k} X_n \leq x\right) P(N = k)$$

$$= P(N = 0) + \sum_{k=1}^{\infty} P\left(\sum_{n=1}^{k} X_n \leq x\right) P(N = k). \tag{5.1}$$

We halt here and note that if $P(N = 0) = p_0 > 0$, or to put it in words, if during the year there is a positive probability that no risk events occur, or there is a positive probability of experiencing no losses during the year, in order to determine the density of total loss we shall have to condition out the no-loss event, and we eventually concentrate on

$$P(S \leq x; N > 0) = \sum_{k=1}^{\infty} P\left(\sum_{n=1}^{k} X_n \leq x\right) P(N = k).$$

To avoid trivial situations, we shall also suppose that $P(X_n > 0) = 1$. This detail will be of importance when we set out to compute the Laplace transform of the density of losses.

From (5.1) we can see the nature of the challenge: First, we must find the distribution of losses for each n, that is given that n risk events took place, and then we must sum the resulting series. In principle, the first step is 'standard procedure' under the IID hypothesis placed upon the X_n. If we denote by $F_X(x)$ the common distribution function of the X_n, then for any $k > 1$ and any $x \geq 0$ we have

$$F_X^{*k}(x) = P\left(\sum_{n=1}^{k} X_n \leq x\right) = \int_0^x F_X^{*(k-1)}(x-y) dF_X(y), \tag{5.2}$$

where $F_X^{*k}(x)$ denotes the convolution of $F_X(x)$ with itself k times. This generic result will be used only in the discretized versions of the X_n, which are necessary for numerical computations. For example, if the individual losses were modeled as integer multiples of a given amount δ, and for uniformity of notation we put $P(X_1 = n\delta) = f_X(n)$, in this case (5.2) becomes

$$P(S_k = n) = f_X^{*k}(n) = \sum_{m=1}^{n} f_X^{*(k-1)}(n-m) f_X(m). \tag{5.3}$$

Thus, the first step in the computation of the aggregate loss distribution (5.1), consists of computing these sums. If we suppose that $F_X(x)$ has a density $f_X(x)$, then differentiating (5.2) we obtain the following recursive relationship between densities for different numbers of risk events:

$$f_X^{*k}(x) = \int_0^x f_X^{*(k-1)}(x-y) f_X(y) dy, \tag{5.4}$$

where $f_X^{*k}(x)$ is the density of $F_X^{*k}(x)$, or the convolution of $f_X(x)$ with itself k times. This is the case with which we shall be concerned from Chapter 6 on.

Notice now that if we set $p_k^0 \equiv p_k/P(N > 0)$, then

$$P(S \leq x \mid N > 0) = \sum_{k=1}^{\infty} P\left(\sum_{n=1}^{k} X_n \leq x\right) p_k^0.$$

We can now gather the previous comments as Lemma 5.1.

Lemma 5.1. *Under the standard assumptions of the model, suppose furthermore that the X_k are continuous with common density $f_X(x)$, then, given that losses occur, S has a density $f_S(x)$ given by*

$$f_S(x) = \frac{d}{dx} P(S \leq x \mid N > 0) = \sum_{k=1}^{\infty} f_{S_k}(x) p_k^0 = \sum_{k=1}^{\infty} f_X^{(k)}(x) p_k^0.$$

This means, for example, that the density obtained from empirical data should be thought of as a density conditioned on the occurrence of losses. The lemma will play a role below when we relate the computation of the Laplace of the loss distribution to the empirical losses. This also shows that in order to compute the density f_S from the building blocks, a lot of work is involved, unless the model is coincidentally simple.

5.2 Analytical techniques: The moment generating function

These techniques have been developed because of their usefulness in many branches of applied mathematics. To begin with, is it usually impossible, except in very simple cases, to use (5.2) or (5.4) to explicitly compute the necessary convolutions to obtain the distribution or the density of the total loss S.

The simplest example that comes to mind is the case in which the individual losses follow a Bernoulli distribution with $X_n = 1$ (or any number D) with probability $P(X_n = 1) = q$ or $X_n = 0$ with $P(X_n = 0) = 1 - q$. In this case $\sum_{n=1}^{k} X_k$ has binomial distribution $B(k, q)$ and the series (5.2) can be summed explicitly for many frequency distributions.

A technique that is extensively used for analytic computations is that of the Laplace transform. It exploits the compound nature of the random variables quite nicely. It takes a simple computation to verify that

$$\phi_{S_k}(\alpha) = E[e^{-\alpha S_k}] = (E[e^{-\alpha X_1}])^k.$$

For the simple example just mentioned, $E[e^{-\alpha X_1}] = 1 - q + qe^{-\alpha}$ and therefore $E[e^{-\alpha S_k}] = (1 - q + qe^{-\alpha})^k$ from which the distribution of S_n can be readily obtained. To continue, the Laplace transform of the total severity is given by:

Lemma 5.2. *With the notations introduced above we have*

$$\psi_S(\alpha) = E[e^{-\alpha S}] = \sum_{k=0}^{\infty} E[e^{-\alpha S_k}] p_k = \sum_{k=1}^{\infty} (E[e^{-\alpha X_1}])^k p_k = G_N(E[e^{-\alpha X_1}]) \quad (5.5)$$

where

$$G_N(z) := \sum_{n=0}^{\infty} z^n P(N = n), \quad \text{for } |z| < 1.$$

We shall come back to this theme in Chapter 6. For the time being, note that for $z = e^{-\alpha}$ with $\alpha > 0$, $G_N(e^{-\alpha}) = \psi_N(\alpha)$ is the Laplace transform of the a discrete distribution.

5.2.1 Simple examples

Let us consider some very simple examples in which the Laplace transform of the compound loss can be computed and its inverse can be identified. These simple examples complement the results from Chapters 2 and 3.

The frequency N has geometric distribution
In this case $p_k = P(N = k) = p(1 - p)^k$ for $k \geq 0$. Therefore

$$\psi_S(\alpha) = E[e^{-\alpha S}] = p\left[1 + \frac{1}{1 - pE[e^{-\alpha X_1}]}\right].$$

For the Bernoulli individual losses,

$$\psi_S(\alpha) = p\left[1 + \frac{1}{1 - p(1 - q + qe^{-\alpha})}\right].$$

The frequency N is binomial B(M, p)

$$\psi_S(\alpha) = E[e^{-\alpha S}] = \sum_{k=0}^{M} \binom{M}{k} (E[e^{-\alpha X_1}])^k = [1 + pE[e^{-\alpha X_1}]]^M.$$

The frequency N is Poisson (λ)

Since we shall be using this model for the frequency of events, we shall explore this example in some detail.

Now $p_k = \lambda^k e^{-\lambda}/k!$, and the sums mentioned above can be computed in closed form, yielding

$$\psi_S(\alpha) = E[e^{-\alpha S}] = \sum_{k=0}^{\infty} \frac{\lambda^n e^{-\lambda}}{k!} (E[e^{-\alpha X_1}])^k = e^{\lambda(E[e^{-\alpha X_1}]-1)}.$$

Some computations with this example are straightforward. For example

$$E[N] = V(N) = E[(N - E[N])^3] = \lambda.$$

Therefore, if we put $\mu = E[X]$,

$$E[S] = \lambda\mu, \quad \sigma_S^2 = \lambda\mu^2 + \lambda\sigma_X^2 = \lambda E[X^2].$$

In particular, from (5.9)

$$E[(S - E[S])^3] = \lambda E[X_1]^3 + 3\lambda\mu\sigma_{X_1}^2 + \lambda E[(X_1 - E[X_1])^3]$$

we obtain that

$$E[(S - E[S])^3] = \lambda E[X_1^3].$$

Thus, the coefficient of asymmetry of the accumulated loss is given by

$$\frac{E[(S - E[S])^3]}{\sigma_S^3} = \frac{E[X_1^3]}{\sqrt{\lambda E[X_1^2]^3}}.$$

Note that this quantity is always positive (X_1 is a positive random variable) regardless of the sign of the asymmetry of X_1.

N is Poisson with a random parameter

Let us build upon the previous model and suppose now that the parameter of the frequency is random, but that the frequency still is Poisson, conditional on the value of the parameter. We shall see how this allows us to modify the skewness in the total losses.

Suppose that the intensity parameter λ is given by the realization of a positive random variable Λ with distribution $H(d\lambda)$ such that

$$P(N = k \mid \Lambda = \lambda) = \frac{\lambda^k}{k!} e^{-\lambda}.$$

In this case the results obtained above can be written as

$$E[S \mid \Lambda = \lambda] = \lambda\mu, \quad V(S \mid \Lambda = \lambda) = \lambda\mu^2 + \lambda\sigma_X^2 = \lambda E[X^2]$$

and, in particular

$$E[(S - E[S])^3 \mid \Lambda = \lambda] = \lambda E[X^3].$$

Now, integrating with respect to $H(d\lambda)$ we obtain

$$E[S] = E[\Lambda]\mu \quad \sigma_S^2 = V(\Lambda)\mu^2 + E[\Lambda]\sigma_X^2$$

and furthermore,

$$E[(S - E[S])^3] = E[(\Lambda - E[\Lambda])^3]\mu^3 + 3\sigma_\Lambda^2 \mu E[X^2] + E[\Lambda]E[X^3].$$

Clearly, if the asymmetry in the distribution of Λ is large and negative, then S can be skewed to the left.

It is also easy to verify that

$$\psi_S(\alpha) = E[E[e^{-\alpha S} \mid \Lambda]] = E[e^{\Lambda(\phi_X(\alpha)-1)}] = G_\Lambda(\phi_X(\alpha) - 1).$$

We leave it up to the reader to verify that the moments of S are obtained by differentiating with respect to α at $\alpha = 0$.

Gamma type individual losses

To complete this short list of simple examples, we suppose that $X_n \sim \Gamma(a, b)$, that is, $f_{X_1}(x) = b^a x^{(a-1)} e^{-bx}/\Gamma(a)$. Observe that, in the particular case when $a = 1$, the gamma distribution becomes the exponential distribution. In the generic gamma case,

$$E[e^{-\alpha X_1}] = \left(\frac{b}{b+\alpha}\right)^a \Rightarrow E[e^{-\alpha S_k}] = E[e^{-\alpha X_1}]^k = \left(\frac{b}{b+\alpha}\right)^{ak}.$$

Or to put it in words, the sum of k random variables distributed as $\Gamma(a, b)$ is distributed according to a $\Gamma(ka, b)$ law.

For any frequency distribution we have

$$\psi_S(\alpha) = \sum_{k=0}^{\infty} p_k \left(\frac{b}{b+\alpha}\right)^{ak},$$

which can be explicitly summed for the examples that we considered above. But more important than that, according to Lemma 5.1,

$$f_S(x) = \frac{d}{dx} P(S \le x \mid N > 0) = \sum_{k \ge 1} p_k^0 \left(\frac{b^k x^{(ak-1)}}{\Gamma(ak)}\right) e^{-bx}.$$

This sum does not seem to easy to evaluate in closed form except in very special cases, but it is easy to deal with it numerically. Notice now that if, for example, N is geometrical with distribution $p_k = p(1-p)^k$, and $a = 1$ (that is the individual losses are exponentially distributed) then the former sum becomes

$$f_S(x) = \sum_{k\geq 1} p_k^0 \left(\frac{b^k x^{(k-1)}}{(k-1)!} \right) e^{-bx} = pbe^{-pbx}.$$

That is, the accumulated losses are again exponential with a smaller intensity (a larger mean equal to $1/bp$).

5.3 Approximate methods

Many different possible methods exist under this header. Let us explore a few.

5.3.1 The case of computable convolutions

In some instances it may be natural or plain easier to use (5.4) or (5.3) in order to calculate the distribution of aggregate losses. This happens when the density of a finite number of losses can be explicitly calculated or at least reasonably well approximated.

This possibility is important because, even when the density of any finite number of losses can be calculated exactly, the resulting series cannot be explicitly summed. In this case one must determine how many terms of the series to choose to determine the losses in some range to a good approximation, and then deal numerically with the resulting approximate sum.

Simple example
A few lines above we considered a simple example of this situation. Suppose that the frequency of losses N is Poisson with parameter (λ) and that the joint distribution of the X_k is $\Gamma(a,b)$. We have already mentioned that the $X_1 + X_2 + \cdots + X_k$ is distributed according to a $\Gamma(ka,b)$, and in Lemma 5.1 we saw that the density $f_S(x)$ of accumulated losses is given by

$$f_S(x) = \frac{1}{1-e^{-\lambda}} \sum_{n=1}^{\infty} e^{-\lambda} \frac{\lambda^n}{n!} \frac{x^{na-1} b^{na}}{\Gamma(na)} e^{-bx}.$$

As mentioned at the outset, the pending issue is now how to chose the range within which we want to obtain a good approximation to that sum. The answer depends on what we want to do. For example, we may just want to plot the values of the function up to a certain range, or we may want to compute some expected values, or we may want to compute some quantiles at the tail of the distribution.

5.3.2 Simple approximations to the total loss distribution

Something that uses simple calculations involving the mean, variance, skewness and kurtosis of the frequency of events and of the individual losses and the independence assumptions of the basic model, is to compute the mean, variance and kurtosis of the total severity and use them to approximate the density of aggregate losses. It is simple to verify that

$$\frac{(-1)^k d^k}{d\alpha^k} \ln \psi_S(\alpha)_{\alpha=0} = E[(S - E[S])^k], \quad \text{for } k = 2,3. \tag{5.6}$$

When $k = 1$

$$\frac{-d}{d\alpha} \ln \psi_S(\alpha)_{\alpha=0} = E[N]E[X], \tag{5.7}$$

where $E[X]$ denotes the expected value of a random variable X whose distribution coincides with that of the X_k. When $k = 2$ we have

$$V(S) = V[N]E[X]^2 + E[N]V(X), \tag{5.8}$$

in which we use $V(Y)$ to denote the variance of a random variable Y. Another identity that we use below is that for $k = 3$

$$E[(S - E[S])^3] = E[(N - E[N])^3]E[X]^3 + 3V(N)E[X]V(X) + E[N]E[(X - E[X])^3]. \tag{5.9}$$

And to finish this preamble, for $k = 4$ it happens that

$$\frac{d^4}{d\alpha^4} \ln \psi_S(\alpha)_{\alpha=0} = E[(S - E[S])^4] - 3V(S). \tag{5.10}$$

We shall now see how moments up to order three of N and X can be used to obtain approximations to the accumulated losses using some simple approximate procedures.

Gaussian approximation

The underlying intuition behind this approximation is that when the expected number of events is large, one can invoke the central limit theorem to obtain an approximation to the density of S. The starting point is the identity

$$P(S \leq x) = P\left(\frac{S - E[S]}{\sigma_S} \leq \frac{x - E[S]}{\sigma_S}\right).$$

From the computations mentioned above, we know that

$$E[S] = E[N]E[X]; \quad \text{and that} \quad V(S) = \sigma_S^2 = \sigma_N^2(E[X])^2 + \sigma_X^2 E[N].$$

We can thus prove that when N is large, the former identity can be rewritten (and it is here that the CLT (central limit theorem) comes in) as

$$P(S \leq x) = \Phi\left(\frac{x - E[S]}{\sigma_S}\right).$$

Here we use the usual $\Phi(x)$ to denote the cumulative distribution of the $N(0,1)$ random variable.

The translated gamma approximation

The Gaussian approximation is simple and based on basic ideas, but it has two handicaps. First, it assigns positive probability to negative losses, and second, it is symmetric. An alternative based on the generic shape of the gamma density consists of supposing the distribution of losses is a translated gamma, and to identify its parameters by a moment matching procedure. That is, we have to determine the parameters (a, b) of a $\Gamma(a, b)$ distributed random variable Y, and a positive number k such that the three first moments of the random variable $k + Y$ coincide with those of the random variable S.

If we denote by γ the coefficient of asymmetry of S, the identities that determine the parameters of $k + Y$ are

$$E(k + Y) = k + \frac{a}{b} = E[S]$$

$$V(k + Y) = \frac{a}{b^2} = \sigma_S$$

$$\frac{E[(k + Y - E(k + Y))^3]}{[V(k + Y)]^{3/2}} = \frac{2}{\sqrt{a}} = \gamma.$$

It is important to keep in mind that this computation makes sense only when $\beta > 0$. Once k, a and b are determined, one can use the translated gamma $k + Y$ to carry out all sorts of computations. A complementary approximation occurs when $2a$ is an integer, because in this case $2aY \sim \chi^2_{2a}$.

5.3.3 Edgeworth approximation

Let us now consider the standardized version of the aggregate loss S, that is, consider

$$Z = \frac{S - E[S]}{\sigma_S}.$$

The Taylor expansion of $\ln \phi_Z(t)$ begins as follows:

$$\ln \phi_Z(\alpha) = a_0 - a_1\alpha + a_2\frac{\alpha^2}{2} - a_3\frac{\alpha^3}{6} + a_4\frac{\alpha^4}{24} \ldots,$$

where $a_k = (-1)^k d^k \ln \phi_Z(\alpha)/d\alpha^k|_{\alpha=0}$. As the variable Z is standardized, clearly

$$a_0 = 0, \quad a_1 = \mathrm{E}[Z] = 0, \quad a_2 = \sigma_Z^2 = 1, \quad a_3 = \mathrm{E}[Z^3],$$

and the fourth moment is

$$a_4 = \mathrm{E}[Z^4] - 3 = \frac{\mathrm{E}[(S - \mathrm{E}[S])^4]}{\sigma_S^4} - 3.$$

Let us now note that if we truncate the series at the fourth term, we obtain

$$\phi_Z(\alpha) \approx e^{\alpha^2/2} e^{-a_3 \alpha^3/6 + a_4 \alpha^4/24}.$$

If we now expand the second exponential and retain the terms up to the sixth order in α,

$$\phi_Z(\alpha) \approx e^{\alpha^2/2}\left(1 - a_3 \frac{\alpha^3}{6} + a_4 \frac{\alpha^4}{24} + a_3^2 \frac{\alpha^6}{72}\right).$$

If we replace α by $-\alpha$ we obtain the usual moment generating function

$$M_Z(\alpha) \approx e^{\alpha^2/2}\left(1 - a_3 \frac{\alpha^3}{6} + a_4 \frac{\alpha^4}{24} + a_3^2 \frac{\alpha^6}{72}\right).$$

We know that

$$e^{\alpha^2/2} = \int_\mathbb{R} e^{-x\alpha} \frac{e^{-x^2/2}}{\sqrt{2\pi}} dx = \int_\mathbb{R} e^{-\alpha x} \frac{d\Phi(x)}{dx} dx,$$

where $\Phi(x)$ is cumulative distribution function of an $N(0,1)$ random variable, therefore

$$(-\alpha)^n e^{\alpha^2/2} = \int_\mathbb{R} \frac{d^n e^{-\alpha x}}{dx^n} \frac{e^{-x^2/2}}{\sqrt{2\pi}} dx = (-1)^n \int_\mathbb{R} e^{-\alpha x} \frac{d^{n+1}\Phi(x)}{dx^{n+1}} dx,$$

where the last step follows from an integration by parts. That is,

$$\phi_Z(\alpha) \approx \int_\mathbb{R} e^{-\alpha x}\left(d\Phi(x) - \frac{a_3}{6} d\Phi^{(3)}(x) + \frac{a_4}{24} d\Phi^{(4)}(x) + \frac{a_3^2}{72} d\Phi^{(6)}(x)\right).$$

As both sides of the equation are Laplace transforms, it follows that

$$P(Z \leq x) \approx \Phi(x) - \frac{a_3}{6} \Phi^{(3)}(x) + \frac{a_4}{24} \Phi^{(4)}(x) + \frac{a_3^2}{72} \Phi^{(6)}(x).$$

Now, undoing the standardization $Z \to (S - \mu_S)/\sigma_S$, we obtain

$$P(S \leq x) = P\left(\frac{S - \mu_S}{\sigma_S} \leq \frac{x - \mu_S}{\sigma_S}\right)$$

$$\approx \Phi\left(\frac{x - \mu_S}{\sigma_S}\right) - \frac{a_3}{6}\Phi^{(3)}\left(\frac{x - \mu_S}{\sigma_S}\right) + \frac{a_4}{24}\Phi^{(4)}\left(\frac{x - \mu_S}{\sigma_S}\right) + \frac{a_3^2}{72}\Phi^{(6)}\left(\frac{x - \mu_S}{\sigma_S}\right). \quad (5.11)$$

A very common situation in practice consists of the case in which S is a compound variable, with frequency N being Poisson(λ). In this case, it is simple to compute the coefficients a_k. It takes a simple computation to verify that

$$\mu_S = \lambda \mu_X, \quad \sigma_S = \sqrt{\lambda \mu_X(2)}, \quad \text{and in general} \quad a_k = \frac{\lambda \mu_X(k)}{(\lambda \mu_X(2))^{k/2}},$$

where we have put $\mu_X(k) = E[X^k]$. In this case, (5.11) becomes

$$P(S \leq x) \approx \Phi\left(\frac{x - \mu_S}{\sqrt{\lambda \mu_X(2)}}\right) - \frac{\mu_X(3)}{6\sqrt{\lambda \mu_X(2)^3}}\Phi^{(3)}\left(\frac{x - \mu_S}{\sqrt{\lambda \mu_X(2)}}\right)$$

$$+ \frac{\mu_X(4)}{24 \lambda \mu_X(2)^2}\Phi^{(4)}\left(\frac{x - \mu_S}{\sqrt{\lambda \mu_X(2)}}\right) + \frac{\mu_X(3)^2}{72 \mu_X(2)^2}\Phi^{(6)}\left(\frac{x - \mu_S}{\sqrt{\lambda \mu_X(2)}}\right).$$

5.4 Numerical techniques

In this section we review some direct numerical techniques for computing the distribution of losses when models for the frequency of losses and for the individual losses are at our disposal, or if not, when empirical data about each block is available. We shall examine how to use that information to compute the distribution of S using (5.4) or from (5.3) plus a direct numerical procedure.

5.4.1 Calculations starting from empirical or simulated data

The procedure described in (5.3) to compute the distribution of losses when the blocks are discrete can be used to deal with empirical or simulated data, for in this case the convolutions have to be carried out numerically.

In both cases we shall suppose that the postanalysis data is of the following two types. First, the relative frequency of events is presented as a distribution with empirical frequencies $\hat{p}_n$ describing the probability that n events occurred during a certain number of years. We furthermore suppose that the number of events is finite, $0 \leq n \leq M$ where M is the maximum number of observed events.

Second, we shall suppose that the frequency of individual losses has been organized in cells, chosen as multiples of a monetary unit δ and we interpret the histogram $P(X = k\delta) = \hat{f}_X(k)$ as the probability of having observed a loss of size $k\delta$.

Observe that if losses occur, then it is necessary that $X > 0$. We must therefore keep in mind that if the empirical (or simulated range of X) is $\delta, 2\delta, \ldots, N\delta$, then the range of the aggregate sum $S_n = X_1 + \cdots + X_n$ is $n\delta \leq S_n \leq nN\delta$. This detail has to be kept in mind when computing the distribution of S_n, that is, when carrying out the computation of

$$\widehat{f}_X^{*n}(k) = \sum_{j=1}^{k} \widehat{f}_X^{*(n-1)}(k-j)\widehat{f}_X(j); \quad n \leq k \leq nN.$$

Once we have computed the $\widehat{f}_X^{*n}$, we can compute their cumulative distribution as follows:

$$\widehat{F}_X^{*n}(x) = \sum_{k\delta \leq x} \widehat{f}_X^{*n}(k),$$

and after that we can estimate the total distribution of losses by

$$\widehat{F}_S(x) = \widehat{p}_0 + \sum_{n=1}^{M} \widehat{p}_n \widehat{F}_X^{*n}(x). \tag{5.12}$$

Comment. When the losses are modeled by a discrete random variable, and no analytical procedure can be implemented, we have to resort to something like what we described before. If the individual losses are modeled by an unbounded random variable, then there will be no upper limit to the partial sums and the numerical issues must include an analysis of the choice of a truncation point.

5.4.2 Recurrence relations

An important technique, less demanding computationally and very much used in the insurance business, is based on the following recurrence methodology.

Suppose that the frequency of events is of the $(a, b, 0)$ class. Recall that this means that for $n \geq 1$ the recurrence relation $p_n = (a+b/n)p_{n-1}$ holds, where p_n is the probability of observing n events. Suppose furthermore that the individual losses occur in multiples of δ, then the aggregate losses take values that are multiples of δ and the probabilities $p_S(n) = P(S = n\delta)$ satisfy the following recurrence relation:

$$p_S(n) = \sum_{k=1}^{n}\left(a + \frac{bk}{n}\right)\frac{p_X(k)p_S(n-k)}{1 - ap_X(0)}; \quad n \geq 1. \tag{5.13}$$

That is, starting from some given $P(S = 0) = P(N = 0)$ and having computed $p_X(n)$ for $n \geq 0$, by iterating the recurrence (5.13) we can find the loss distribution. For more about this, see [78] and [100].

5.4.3 Further numerical issues

We have already mentioned several times that most of the time we may have to resort to numerical procedures to calculate sums like (5.3) and (5.4), that is to evaluate series like

$$F_S(x) = p_0 + \sum_{n=1}^{\infty} p_n F_X^{*n}(x).$$

There are two complementary issues here. First, there is the choice of the number of terms of the series to be summed, and second the range $[0, L]$ in which we want $F_S(x)$ to be determined. To answer the first question is easy due to the uniform boundedness of the $F_X^{*n}(x)$. To state it formally, consider:

Lemma 5.3. *For every $\epsilon > 0$ there exists an integer $K > 0$ such that*

$$F_S(x) - \sum_{n=0}^{\infty} p_n F_X^{*n}(x) < \epsilon$$

uniformly in $x \in [0, \infty)$.

The proof is simple. Since $\sum_{n \geq 0} p_n = 1$, there is an M such that $\sum_{n \geq M+1} p_n < \epsilon$, from which the conclusion emerges. It is implicit in the proof that the F_X^{*n} are known.

When we are forced to compute the F_X^{*n}, what to do depends on the starting point. If we have a reliable model for $F_X(x)$, then we can compute the iterated integrals either using integral transforms, or numerically using some discretization procedure. If all we have is empirical data, and cannot reduce to the previous case by fitting a distribution to the data, then we have to proceed numerically.

We shall consider all these alternatives below, but for the time being, let us begin by recalling some direct approximations that have been used in actual practice.

5.5 Numerical examples

In this section we present a collection of detailed examples. It is a vital section for students that must come up with this material through worked examples. The examples covered are part of the daily life of a risk analyst confronted with small sized empirical data.

Example 1. Consider to begin with the following simple example. Let us suppose that the average monthly frequency of events of a certain type is eight events per month and that its standard deviation is of three events per month. Suppose as well that the average loss per event is 150 000, with a standard deviation of 30 000 in some monetary units. If we apply (5.6) we obtain that

$$E[S] = \mu_S = 8 \times 150\,000\, \$ = 12 \times 10^5,$$
$$V(S) = 9 \times (15 \times 10^5)^2 + 8 \times (3 \times 10^4)^2 \approx (3 \times 15)^2 \times 10^{10} \Rightarrow \sigma_S = 45 \times 10^5.$$

Let us suppose furthermore that the distribution that better fits this data is the lognormal with parameters (μ, σ), and we want to determine its parameters from the data. We know that

$$E[S] = e^{\mu+\sigma^2/2} \quad \text{and that} \quad E[S^2] = e^{2\mu+2\sigma^2}.$$

With the numbers given above, we have to solve

$$\mu + \sigma^2/2 = 5 + \log_{10}(12)$$
$$\mu + \sigma^2 = 5 + \log_{10}(54)$$

all of which yields $\mu = 5.51$, $\sigma = 1.07$.

Solving for (μ, σ) we know that the aggregate loss is given by $S = \exp(\mu + \sigma Z)$ with $Z \sim N(0, 1)$. Once we have reached this point, we can answer questions like: What is the probability that the losses exceed the mean by a certain number of standard deviations? That is, we want to compute $P(S > \mu_S + a\sigma_S)$, or perhaps quantities like $E[S \mid S > v]$ where v is some large quantile. We leave it as an exercise for the reader to answer a few such questions.

Example 2. In the next example we illustrate several aspects of the routine that should be carried out in a real situation. We want to illustrate how equation (5.1) is used as a technique of loss aggregation. Suppose that the individual losses in a small business, in thousands of some monetary unit, are given in Table 5.1. Suppose as well that the frequency of the losses is as in Table 5.2 and the recorded frequencies are listed in Table 5.2.

Table 5.1: Loss distribution.

x	$p_X(x)$	$F_X(x)$
1	0.01024	0.01024
2	0.07680	0.08704
3	0.23040	0.31744
4	0.34560	0.66304
5	0.25920	0.92224
6	0.07776	1.00000

Table 5.2: Frequency distribution.

k	0	1	2	3	4
p_k	0.05	0.10	0.20	0.25	0.4	1

In Figure 5.1 we display the values of $p_X(x)$ according to Table 5.1 and the histogram of the frequency of events shown in Table 5.2.

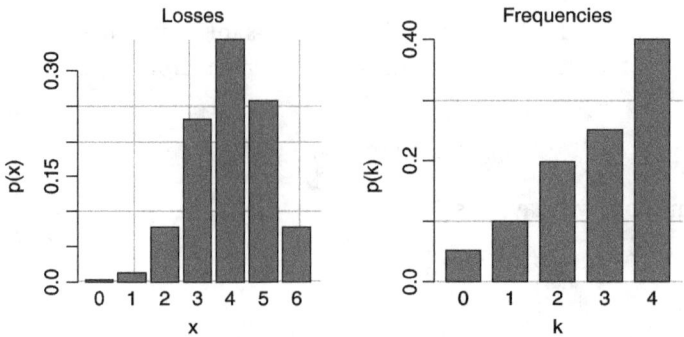

Figure 5.1: Histograms of data in Tables 5.1 (left panel) and 5.2 (right panel).

From the data in Tables 5.1 and 5.2 we want to know the probability that the aggregate loss is larger or smaller than 10 (in units of 1000) as well as the expected aggregate loss $E[S]$ and its variance $V[S]$.

Since $S = X_1 + X_2 + \cdots + X_N$ (recall that $S = 0$ whenever $N = 0$), the probability $P(S \leq 10)$ is computed like

$$P(S \leq x) = F_S(x) = \sum_{k=0}^{\infty} p_k F_X^{*k}(x),$$

with $x = 10$ (corresponding to 10 000) and $p_k = P(N = k)$ being given in Table 5.2.
When $k = 0$, the value of $F_X^{*0}(x)$ is obtained as follows:

$$F_X^{*0}(x) = \begin{cases} 0, & \text{if } x < 0 \\ 1, & \text{if } x \geq 0. \end{cases}$$

For $k = 1$ we have $F_X^{*1}(x) = F_X(x)$. In Table 5.3 we show the values of $F^{*0}(x)$ and $F^{*1}(x)$. Nothing new so far.

Table 5.3: $F_X^{*0}(x)$ and $F_X^{*1}(x)$.

x	0	1	2	3	4	5
$F_X^{*0}(x)$	1.00000	1.00000	1.00000	1.00000	1.00000	1.00000
$F_X^{*1}(x)$	0	0.01024	0.08704	0.31744	0.66304	0.92224
x	6	7	8	9	10	11
$F_X^{*0}(x)$	1.00000	1.00000	1.0000	1.00000	1.00000	1.0000
$F_X^{*1}(x)$	1.00000	1.00000	1.00000	1.00000	1.00000	1.00000

Since X is discrete and its values are consecutive integers, from the knowledge of $p_X(x)$ for $x = \{0, 1, 2, \ldots, 10\}$, we can obtain the value $F_X^{*k}(x)$ for $k = \{2, 3, 4\}$ by simple iteration, that is

$$F_X^{*k}(x) = \sum_{y=0}^{x} F_X^{*(k-1)}(x-y)p_X(y).$$

The cumulative distribution function $F_X^{*k}(x)$ for losses up to $\$11.000$ are collected in Table 5.4, along with the values of $F_S(x)$.

Table 5.4: $F^{*k}(x)$ and $F_S(x)$.

x	$F_X^{*0}(x)$	$F_X^{*1}(x)$	$F_X^{*2}(x)$	$F_X^{*3}(x)$	$F_X^{*4}(x)$	$F_S(x)$
0	1	0.00000	0.00000	0.00000	0.00000	0.05000
1	1	0.01024	0.00000	0.00000	0.00000	0.05102
2	1	0.08704	0.00010	0.00000	0.00000	0.05872
3	1	0.31744	0.00168	0.00000	0.00000	0.08208
4	1	0.66304	0.01229	0.00003	0.00000	0.11877
5	1	0.92224	0.05476	0.00028	0.00000	0.15325
6	1	1.00000	0.16624	0.00193	0.00001	0.18373
7	1	1.00000	0.36690	0.00935	0.00005	0.22574
8	1	1.00000	0.61772	0.03383	0.00032	0.28213
9	1	1.00000	0.83271	0.09505	0.00161	0.34095
10	1	1.00000	0.95364	0.21310	0.00647	0.39659
11	1	1.00000	0.99395	0.39019	0.02103	0.45475
p_n	0.05	0.10	0.20	0.25	0.40	–

That is, the value of $P(S \leq 10) = F_S(10) = \sum_{k=0}^{4} p_k F_X^{*k}(10)$ happens to be 0.39659, as can be seen from Table 5.4.

Another possible way to compute $P(S \leq 10)$ is to directly compute the probability function $P(S = k)$ of S, and then simply use it to calculate whatever we need. In this case (5.1) has to be modified a bit. To unify notations we shall write $P(X = x)$ as $f_X(x)$, $P(S_k = x) = f_X^{*k}(x)$. Then clearly

$$f_S(x) = \sum_{k=0}^{\infty} p_k f_X^{*k}(x),$$

where now

$$f_X^{*0}(x) = \begin{cases} 1, & \text{whenever } x = 0 \\ 0, & \text{for } x \neq 0. \end{cases}$$

For $k = 1$ clearly

$$f_X^{*1}(x) = \begin{cases} f_X(x), & \text{when } 1 \le x \le M \\ 0, & \text{otherwise,} \end{cases}$$

where in our numerical example $M = 6$. And for $k > 1$, since the range of S_k is $[k, kM]$,

$$f_X^{*k}(x) = \begin{cases} \sum_{y=1}^{x} f_X^{*k}(x-y) f_X^{*1}(y), & \text{for } k \le x \le kM \\ 0, & \text{otherwise} \end{cases}$$

$$f_X^{*k}(x) = \sum_{y=0}^{x} f_X^{*(k-1)}(x-y) f_X(y); \quad k = 2, 3, 4 \ldots$$

The results of the computations are displayed in Table 5.5. In the rightmost column we display the values of $P(S = x) \equiv f_S(x)$ for $x = 0, \ldots 10$ instead of the values of the cumulative distribution.

Table 5.5: Values of $f^{*k}(x)$ and $f_S(x)$.

x	$f_X^{*0}(x)$	$f_X^{*1}(x)$	$f_X^{*2}(x)$	$f_X^{*3}(x)$	$f_X^{*4}(x)$	$f_S(x)$
0	1	0.00000	0.00000	0.00000	0.00000	0.05000
1	0	0.01024	0.00000	0.00000	0.00000	0.00102
2	0	0.07680	0.00010	0.00000	0.00000	0.00770
3	0	0.23040	0.00157	0.00000	0.00000	0.02335
4	0	0.34560	0.01062	0.00002	0.00000	0.03669
5	0	0.25920	0.04247	0.00025	0.00000	0.03448
6	0	0.07776	0.11148	0.00165	0.00000	0.03049
7	0	0.00000	0.20066	0.00742	0.00004	0.04200
8	0	0.00000	0.25082	0.02449	0.00027	0.05639
9	0	0.00000	0.21499	0.06121	0.00129	0.05882
10	0	0.00000	0.12093	0.11806	0.00485	0.05564
11	0	0.00000	0.04031	0.17708	0.01456	0.05816
p_n	0.05	0.10	0.20	0.25	0.40	–

From the results in the last column we see that $P(S \le 10) = \sum_{s=0}^{10} f_S(s)$ is 0.39659, which coincides with the result determined above. The first two moments of S are easy to estimate as well. Note that $E[N] = 2.85$, $\sigma(N) = 1.1945$, $E[X] = 3.9972$, and $\sigma(X) = 1.142926$. Therefore,

$$E(S) = E(N)E(X) = 11.3972$$

$$V(S) = E(N)V(X) + (E(X))^2 V(N) = 22.8082 \Rightarrow \sigma(S) = 4.7770.$$

These will be used as inputs in the following example, in which we compare the exact calculation carried out above with the procedures examined in the first example.

Example 3. Consider a small financial institution that has some data, but not in very large quantities, so that its technicians cannot infer individual loss and frequency distributions with high confidence.

In this case, the analyst, before attempting computations like those in the previous example, can try using approximate methods to extract interesting information for the risk managers.

One possibility may consist of using simple analytical approximation techniques to estimate the probability that losses are smaller than some threshold, say 10, in some appropriate units.

Suppose that from the data the technician can confidently infer a few moments of N and X from data and since she is not carrying out the convolutions, she computes a few moments instead to answer questions that interests risk management. For example, what is $P(S < 10)$ if we regard 1000 as the unit of measurement? She will answer it using the normal, gamma, and lognormal approximations. We shall suppose that the only data available is that reported at the end of the previous example. With that we compute the first two moments to obtain

$$P(S \leq 10) \approx P\left(\frac{S - E(S)}{\sqrt{\text{Var}(S)}} \leq \frac{10 - 11.3972}{4.7758}\right) = P(Z \leq -0.2925) = 0.3850.$$

Clearly, in this case, the Gaussian approximation underestimates the probability of loss. Supposedly the Gaussian approximation is to be trusted when $E[N]$ is large, and perhaps this is the cause of the poor approximation in our example.

The risk manager may mistrust the previous approximation and decides to examine a lognormal and a translated gamma for the sake of comparison. To determine the parameters of the lognormal we must solve for (μ, σ) in

$$E(S) = \exp\left(\mu + \frac{1}{2}\sigma^2\right) \quad \text{and} \quad E(S^2) = V(S) + E[X]^2 = \exp(2\mu + 2\sigma^2).$$

The solution of this system of equation is simple and it yields: $\mu = 2.3115$ and $\sigma = 0.2442$:

$$P(S \leq 10) = \Phi\left[\frac{\ln(10) - 2.3115}{0.2442}\right] \approx 0.4854,$$

which overestimates the probability. Perhaps exponentiating a normal makes the tail too heavy. However, for this approximation one only needs the mean and the variance of the total loss, quantities which are easily obtainable from he empirical data. The other approximate method, described in Section 5.3.2, is the translated gamma approximation. Recall that this consists of replacing S with $k+Y$, where $Y \sim \Gamma(a, b)$, and the coefficients k, a, b have to be determined to match the mean, variance and the asymmetry coefficients of S. We explained there how to compute these from the data about the frequency of events N and the individual losses X. From the information provided in Example 2 we can easily do that. Explicitly,

$$E(S) = E(N)E(X) = \mu = 11.3972$$

$$V(S) = E(N)V(X) + V(N)(E(X))^2 = \sigma^2 = 28.8081$$

$$\frac{E[(S - E(S))^3]}{[V(S)]^{3/2}} = \gamma = 1.0178.$$

Recall as well that in terms of the block data,

$$E[(S - E(S))^3] = E(N)E[(X - E[X])^3] + 3V(N)E(X)V(X) + E[(N - E[N])^3](E(X))^3.$$

To determine the coefficients we have to solve the system

$$\mu = k + \frac{\alpha}{\beta}$$

$$\sigma^2 = \frac{\alpha}{\beta^2}$$

$$\gamma = \frac{2}{\sqrt{\alpha}}.$$

Carrying out the computations we obtain $\gamma = 1.0178$, and we have already seen that $E[S] = 11.4$ and $V(S) = 35.953$. From this we obtain the coefficients of the gamma as $\alpha = 3.861$ and $\beta = 0.4145$ and for the translation coefficient we get $k = 2.008$.

Now to use this to estimate $P(S \leq 10)$ we proceed as follows:

$$P(S \leq 10) = P(k + Y \leq 10) = P(Y \leq 7.9992) \approx 1.$$

For the last step use your favorite spreadsheet. Here the overestimation is clear.

Example 4. Consider the following slightly more elaborate version of Example 2. Suppose that the frequency of losses is Poisson with $\lambda = 0.8$, and that individual losses take only three values as indicated in Table 5.6. We want to estimate the probability of not losing more than 8000 in some monetary unit, or that the losses do no exceed 8 in thousands of that unit.

Table 5.6: Individual losses (in thousands).

x	$p_X(x)$	$F_X(x)$
1	0.250	0.250
2	0.375	0.625
3	0.375	1

We know that $P(N = k) = \frac{\lambda^k e^{-\lambda}}{k!}$, for $k = \{0, 1, \ldots\}$. In Table 5.7 we list the probabilities for various values of N. We see from this that considering time lapses in which more than six events occur is not necessary.

Table 5.7: Distribution of frequencies.

k	0	1	2	3	4	5	6	
$p(k)$	0.4493	0.3595	0.1438	0.0383	0.0077	0.0012	0.0002	$\sum_{k=1}^{6} p_k = 1$

With the data on these two tables we can perform computations similar to those performed in Example 2 and we arrive at Table 5.8. There we list the values of the cumulative loss distribution function. From that it is clear that the probability that we are after is 99.3 %.

Table 5.8: Values of $F^{*k}(x)$ and $F_S(x)$.

x	$F_X^{*0}(x)$	$F_X^{*1}(x)$	$F_X^{*2}(x)$	$F_X^{*3}(x)$	$F_X^{*4}(x)$	$F_X^{*5}(x)$	$F_X^{*6}(x)$	$F_S(x)$
0	1	0.000	0.000000	0.0000000	0.00000000	0.0000000000	0.0000000000	0.4493000
1	1	0.250	0.000000	0.0000000	0.00000000	0.0000000000	0.0000000000	0.5391750
2	1	0.625	0.062500	0.0000000	0.00000000	0.0000000000	0.0000000000	0.6829750
3	1	1.000	0.250000	0.0156250	0.00000000	0.0000000000	0.0000000000	0.8453484
4	1	1.000	0.578125	0.0859375	0.00390625	0.0000000000	0.0000000000	0.8952559
5	1	1.000	0.859375	0.2617188	0.02734375	0.0009765625	0.0000000000	0.9426137
6	1	1.000	1.000000	0.5253906	0.10351562	0.0083007812	0.0002441406	0.9735295
7	1	1.000	1.000000	0.7890625	0.26171875	0.0375976562	0.0024414062	0.9848819
8	1	1.000	1.000000	0.9472656	0.49243164	0.1145019531	0.0128784180	0.9928120
p_n	0.4493	0.3595	0.1438	0.0383	0.0077	0.0012	0.0002	–

Example 5. This time we examine the use of the recurrence relations instead of carrying out the convolutions. The underlying model is that considered in the previous example, namely, the individual losses are given in Table 5.6, and the frequency of losses are modeled by a Poisson of intensity $\lambda = 0.8$. Again we want to estimate the probability of losing no more than 8 (in thousands) of the chosen monetary unit.

Since $p_X(0) = 0$, clearly $P(S = 0) = P(N = 0) = e^{-0.8} = 0.4493$ is the probability of no losses taking place. As the frequencies are modeled by a Poisson law, according to the results in Section 5.1.1 we have $a = 0, b = \lambda = 0.8$. For $n > 0$, the recurrence relation is

$$p_S(n) = \sum_{k=1}^{n} \left(a + \frac{bk}{n} \right) \frac{p_X(k) p_S(n-k)}{1 - a p_X(0)}.$$

Therefore,

$$p_S(1) = \lambda p_X(1) p_S(0) = 0.8 \times 0.250 \times 0.4493 = 0.089866$$

$$p_S(2) = \frac{\lambda}{2} p_X(1) p_S(1) + \lambda p_X(2) p_S(0)$$

$$= (0.8) \times (0.25) \times (0.0898) + (0.8) \times (0.375) \times (0.4493) = 0.1437$$

⋮

In Table 5.9 we show the result of carrying out this process a few more times.

Table 5.9: Loss distribution.

S	0	1	2	3	4	5	6	7	8	$\sum_{n=1}^{8} p_S(n)$
p_S	0.4493	0.0898	0.1437	0.1623	0.0499	0.0473	0.0309	0.0011	0.0079	0.9928

As indicated $P(S \leq 8) = \sum_{s=0}^{8} p_S(s) = 0.9928$ as observed in the previous example. The difference between the two cases consists of the number of computations to be performed. The recurrence methodology is quite efficient in this regard.

Example 6. In Table 5.10 we show the individual loss data of a large company (in millions of some monetary unit). Suppose that the losses are arbitrary, but the data is collected in 'bins'. In the table we show the collection bins and their middle points, which play the role of discrete values of the variable. This data will be refined below, when we prepare it for the convolution procedure. The data is plotted in the histogram in Figure 5.2 in which it is apparent that the distribution is bimodal. This can occur for example if there were two sources of risk and the data comes in aggregated across sources.

Table 5.10: Distribution of losses in millions in some monetary unit.

X (in mills.)	M.P	$p_X(x)$	F(x)
(4, 6]	5	0.04802111	0.04802111
(6, 8]	7	0.09498681	0.14300792
(8, 10]	9	0.13403694	0.27704485
(10, 12]	11	0.11029024	0.38733509
(12, 14]	13	0.07176781	0.45910290
(14, 16]	15	0.04010554	0.49920844
(16, 18]	17	0.02638522	0.52559367
(18, 20]	19	0.02849604	0.55408971
(20, 22]	21	0.04010554	0.59419525
(22, 24]	23	0.05593668	0.65013193
(24, 26]	25	0.07387863	0.72401055
(26, 28]	27	0.07968338	0.80369393
(28, 30]	29	0.06965699	0.87335092
(30, 32]	31	0.06385224	0.93720317
(32, 34]	33	0.04538259	0.98258575
(34, 36]	35	0.01741425	1

Suppose furthermore that the frequency of losses during the time lapse of interest is rather simple as specified in Table 5.11. The risk manager needs to estimate the probability that the losses are larger than 50 million.

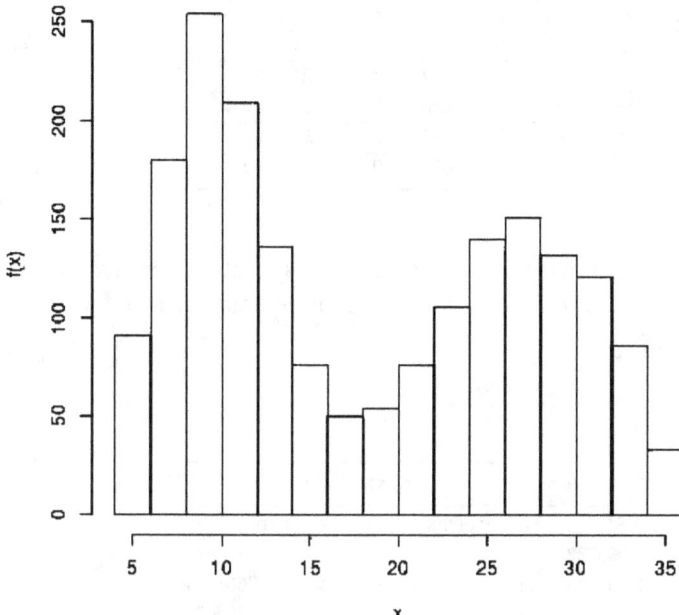

Figure 5.2: Histogram of values in Table 5.10.

Table 5.11: Frequency distribution.

k	0	1	2	3	4	
p_k	0.05	0.10	0.20	0.25	0.40	1

The data collection procedure allows us to apply the techniques employed in the examples described above. In principle, we need to compute

$$P(S \geq x) = 1 - F_S(x) = 1 - \sum_{s=0}^{x} f_S(s),$$

where

$$f_S(x) = \sum_{k=0}^{\infty} p_k f_X^{*k}(x),$$

where again we use $f_X(x)$ instead of $P(X = x)$.

To perform the convolutions, we think of the middle points of the bins in Table 5.10 as the observed losses. Their probabilities obtained from that table are listed below in Table 5.12.

Table 5.12: Discretized individual probability loss distribution (in millions).

x	0	1	2	3	4	5	6	7	8	9	10	11
p_x	0	0	0	0	0	0.048	0	0.095	0	0.1341	0	0.1103
x	12	13	14	15	16	17	18	19	20	21	22	23
p_x	0	0.072	0	0.040	0	0.026	0	0.028	0	0.04	0	0.056
x	24	25	26	27	28	29	30	31	32	33	34	35
p_x	0	0.074	0	0.079	0	0.07	0	0.064	0	0.045	0	0.017

We chose the maximum number of losses to be rather small so that performing the convolutions as in Example 1 is easy. Doing that produces Table 5.13 shown below, in which we compute $P(S = x) \equiv f_S(x)$ from $x = 0$ up to $x = 50$.

Table 5.13: Distribution of S.

x	$f_S(x)$	x	$f_S(x)$	x	$f_S(x)$	x	$f_S(x)$
0	0.0500000000	18	0.0091612125	30	0.0090830433	42	0.0155757319
5	0.0048021110	19	0.0034063732	31	0.0116687812	43	0.0107128249
7	0.0094986810	20	0.0094105720	32	0.0120890097	44	0.0136841256
9	0.0134036940	21	0.0053326462	33	0.0097607880	45	0.0120876057
10	0.0004612054	22	0.0083253624	34	0.0150529158	46	0.0126616145
11	0.0110290240	23	0.0080263849	35	0.0071159614	47	0.0127897115
12	0.0018245488	24	0.0069377001	36	0.0171836458	48	0.013019472
13	0.0071717678	25	0.0110389771	37	0.0060405021	49	0.0127169534
14	0.0043791399	26	0.0062920016	38	0.017954198	50	0.0144556153
15	0.0040382385	27	0.0126125059	39	0.0073014853		
16	0.0072112004	28	0.0069800025	40	0.0173115923		
17	0.0028028036	29	0.0121461642	41	0.0089810793		

Summing up the entries along the even columns of Table 5.13 we obtain $F_S(50) = \sum_{s=0}^{50} f_S(s) = 0.473539$. Therefore, $P(S > 50) = 1 - F_S(50) = 0.526461$, that is the probability that losses exceed 50 million is about 53 %.

5.6 Concluding remarks

There are two main conclusions to be obtained from these examples. First, performing the computations according to basic theory may be lengthy and time consuming, but in some cases is the only thing that we can do. Therefore, there is real need for methods of approximation. It is also clear from the examples that the obvious and simple methods of approximation may yield the wrong estimate of the relevant probabilities. Additionally, these methods need some further theoretical support to determine when and when not to apply them.

In the references at the end of the book other methods are considered along with applications to problems in loss data analysis and operational risk. Also, the examples presented in this section all have their reason of being, and are intended as an appetizer for the methods presented from Chapter 7 on.

6 Laplace transforms and fractional moment problems

This chapter is a detour from the track we were on, but as it is an essential part of the methodology that we are to use to deal with the problem of loss data analysis, we decided to gather the material under one heading. Since the Laplace transform technique is rather important we shall try to weave a line between generality and the particular applications that we have in mind. To make it generic, let us denote by X a positive random variable, and throughout this chapter we suppose that it is continuous and has a density $f_X(x)$, with respect to the usual Lebesgue measure on $[0, \infty)$ unless otherwise specified. Its Laplace transform is defined by

$$\phi(\alpha) = \mathrm{E}[e^{-\alpha X}] = \int_0^\infty e^{-\alpha x} f_X(x) dx. \tag{6.1}$$

The representation of the Laplace transform of f_X as an expected value is useful because in many cases there is information about X that allows us to compute its Laplace transform $\phi(\alpha)$, but not its probability density $f_X(x)$ directly. We have already mentioned the fact that $\phi(\alpha)$ may solve some equation, or that X is a compound random variable and we may know the Laplace transform of its building blocks, or that we may have a sample of values of X from which to compute $\mathrm{E}[\exp(-\alpha X)]$. Our goal is thus to use the knowledge of $\phi(\alpha)$ given by (6.1) to obtain the density f_X.

6.1 Mathematical properties of the Laplace transform and its inverse

The ubiquitousness and importance of the Laplace transform has motivated much research about the properties of the mapping $f(x) \to \phi(\alpha)$ and, especially, the inverse mapping. As a matter of fact, below we shall verify that $f(x) \to \phi(\alpha)$ is continuous and injective, but that its inverse is not continuous. This fact is related to many numerical instability issues. From the mathematical point of view, the $f_X(x)dx = dF_X(x)$ appearing in (6.1) might be a very general distribution. As we are concerned with probability densities on $[0, \infty)$, our aims are much more modest, nevertheless for the statements of some general properties of the Laplace transform throughout much of this section we shall not care too much whether $f \geq 0$ and whether it integrates to 1. By L_p we shall denote the class of Borel measurable functions on $[0, \infty)$ such that $\int |f(x)|^p dx < \infty$.

The first thing to note is that for $f \in L_1$, the transform can actually be extended to the right half complex plane $\{z = \alpha + i\beta \mid \alpha > 0, \beta \in \mathbb{R}\}$. Not only that, but an easy application of Morera's theorem shows that $\phi(z)$ is analytic there and the standard Laplace inversion technique can be invoked to invert it and obtain $f(x)$. The analyticity property will be related below to the determinacy of the fractional moment problem. This is because analytic functions can be recovered from their values at a sequence of points with an accumulation point. The fact that the standard inversion formula consists of a contour integration of an analytic function in the right half complex plane has given rise to a lot of interesting mathematics related to the complex interpolation problem. See for example [53] or [7].

All the comments made above are important to us on several accounts. First, $\phi(a)$ can only be determined numerically at a few, positive, values of a. Second, this usually means that $\phi(a)$ is known up to some error, and as mentioned, the inverse Laplace transformation is not continuous, so our methods will need to be robust. And third, even if we knew $\phi(a)$ exactly at few points and we wanted to use classical inversion formulae, we would still need to extend it to the right half complex plane as an analytic function, which is a major problem in itself. Thus, in order to avoid the complex interpolation problem, we shall propose inversion methods that make use of the data directly.

Let us now examine the issue of continuity of the direct and inverse Laplace transforms. Let $f_1(x), f_2(x)$ both satisfy $|f_i(x)| \leq Me^{ax}$ for some $M, a > 0$, or let both be in $L_1([0, \infty)) \cap L_2([0, \infty))$. Then for $\Re z > a$

$$|\phi_1(z) - \phi_2(z)| = \left| \int_0^\infty (f_1(x) - f_2(x)) e^{-zx} dx \right|$$

$$\leq \int_0^\infty |f_1(x) - f_2(x)| e^{ax} e^{-(z-a)x} dx$$

$$\leq \sup_{x>0} |f_1(x) - f_2(x)| e^{ax} \int_0^\infty e^{-(\Re z - a)x} dx$$

$$< \frac{\varepsilon}{\Re z - a}.$$

Thus if $\sup_{x>0} |f_1(x) - f_2(x)| e^{-ax} \leq \varepsilon$, the difference in Laplace transforms stays small in $\Re z - a > \sigma$ for every $\sigma > 0$.

A similar computation shows that when $\|f_1 - f_2\|_p < \varepsilon$ then $|\phi_1(z) - \phi_2(z)| < M\varepsilon$ uniformly in compact regions of the right half complex plane. That is, the Laplace transform is continuous in many setups.

Consider now the two following examples. First $f_\omega(t) = A \sin(\omega t)$. Its Laplace transform is $\phi_\omega(z) = \frac{\omega A}{z^2 + \omega^2}$ for $\Re z > 0$.

Even though $f_\omega(x)$ has constant amplitude A, which can be small or large, its oscillations get higher and higher as ω increases, but $|\phi_\omega(z)| \leq \frac{A}{\omega}$ (for $\Re z > 0$). Thus small changes in $\phi_\omega(z)$ can produce wildly different reconstructions.

Consider now $f_a(x) = A \frac{ax^{-3/2}}{2\sqrt{\pi}} e^{-a^2/4x}$ together with $\phi_a(a) = Ae^{-a\sqrt{a}}$ with $a > 0$. Notice that for small a, $\phi_a(a)$ stays bounded by A but $f_a(x)$ has a maximum at $x = a^2/6$, of size

cA/a^2, with $c > 0$. Thus even if A were a small number, the hump in $f_a(x)$ can be large if a is small.

Therefore, even though the direct Laplace transform is continuous, its inverse is not, and this is the reason why inverting Laplace transforms is a terribly complicated problem from the numerical point of view. There exists a large body of literature proposing different techniques devised to deal with the problem of inverting Laplace transforms. In the references at the end of the book we mention only a small sample of the large literature on this subject.

6.2 Inversion of the Laplace transform

To better appreciate the methodology that we propose below, let us examine some classical results. The first one is from [23], but consider as well [99] or [30].

Theorem 6.1. *Let $\tilde{\phi}(z)$ be an analytic function in the strip $\alpha < \mathbb{R}(z) < \beta$ of the complex plane. Suppose furthermore that in the narrower strip $\alpha + \delta < \mathbb{R}(z) < \beta - \delta$, for some appropriate δ, we have $\tilde{\phi}(z) \leq k(|z|)$, for some $k \in L_1(0, \infty)$. Then for any $x \in \mathbb{R}$ and fixed y,*

$$\phi(x) = \frac{1}{2\pi i} \int_{y-i\infty}^{y+i\infty} \tilde{\phi}(s) e^{xs} ds \qquad (6.2)$$

is well defined, and furthermore, in the strip $\alpha < \mathbb{R}(z) < \beta$, the following identity holds:

$$\tilde{\phi}(x) = \frac{1}{2\pi i} \int_{-i\infty}^{+i\infty} \phi(x) e^{-xz} dx. \qquad (6.3)$$

The case that is of interest to us corresponds to:

Corollary 6.1. *If in the statement of the theorem, $\beta = \infty$, and $\tilde{\phi}(z)$ is analytic on the right half complex plane $\mathbb{R}(z) > 0$, then*

$$\tilde{\phi}(z) = \int_0^{i\infty} \tilde{\phi}(s) e^{xs} ds, \quad \text{for } \mathbb{R}(z) > 0, \qquad (6.4)$$

and

$$\phi(x) = \frac{1}{2\pi i} \int_{y-i\infty}^{y+i\infty} \tilde{\phi}(s) e^{xs} ds, \qquad (6.5)$$

for y such that all possible poles of $\tilde{\phi}$ have real parts less that y.

Observe that if we wanted to use the inversion formula provided in these statements, and the only information available to us consisted of the value of $\tilde{\phi}$ at a finite number of points on the real axis, first we would have to extend it to an analytic function on vertical strip or to the right of a vertical line in the complex plane, such that all of its poles lie to the left of that line. As previously mentioned, this has given rise a lot of mathematics on the complex interpolation problem, some of which is mentioned in the references.

When studying the Laplace transform of measures on $[0, \infty)$ there exist alternative inversion procedures. Let us begin with:

Definition 6.1. A function $\phi(\lambda)$ defined on $(0, \infty)$ is completely monotone if it possesses derivatives of all orders and

$$(-1)^n \frac{d\phi}{d\lambda}(\lambda); \quad \lambda > 0.$$

An interesting result is contained in:

Lemma 6.1. *If ϕ is completely monotone, and if ψ is positive with completely monotone first derivative, then $\phi(\psi)$ is completely monotone. In particular, $e^{-\psi}$ is completely monotone.*

The following results taken from [34] provide a characterization of Laplace transform of measures as well as an inversion procedure.

Theorem 6.2. *A function $\phi(\lambda)$, defined on $(0, \infty)$ is completely monotone if and only if it is of the form*

$$\phi(\lambda) = \int_0^\infty e^{-x\lambda} dF(x),$$

where $F(x)$ is the cumulative mass function of a (possibly σ-finite) measure on $[0, \infty)$. Furthermore, $F(x)$ can be obtained from $\phi(\lambda)$ as

$$F(x) = \lim_{\lambda \to \infty} \sum_{n \leq \lambda x} \frac{(-\lambda)^n}{n!} \phi^{(n)}(\lambda). \tag{6.6}$$

To close, a potentially useful result is contained in:

Corollary 6.2. *For $\phi(\lambda)$ to be of the form*

$$\phi(\lambda) = \int_0^\infty e^{-\lambda x} f(x) dx$$

it is necessary and sufficient that

$$0 \le \frac{(-a)^n \phi^{(n)}(a)}{n!} \le \frac{C}{a} \quad \text{for all } a > 0$$

for some positive constant C.

Note that when $F(x)$ is the cumulative distribution of a random variable X, having finite moments $E[X^n] = \mu_n$ for all $n \ge 0$, and the domain of analyticity of $\phi(\lambda)$ includes the point $\lambda = 0$, then $(-1)^n \phi^{(n)}(0) = \mu_n$, and the theorem asserts that the Laplace transform and the integer moments determine each other. The extension of this result to fractional moments is the content of Section 6.4 below.

But this is not always the case as the following famous example shows. Let $X = e^Z$ with $Z \sim N(0, 1)$, and let consider the two densities

$$f(x) = e^{-x^2/2}/(x\sqrt{2\pi}) \quad \text{along with} \quad g(x) = f(x)(1 + a\sin(2m\pi(\ln(x)))).$$

Here m is an integer and $|a| < 1$. It only takes a change of variables to verify that

$$\int_0^\infty x^n f(x)dx = e^{n^2/2} = \int_0^\infty x^n g(x)dx.$$

That is, the lognormal density is not determined by its moments. See [65] for more on this issue. Actually, the issue lies with the analyticity of the Laplace transform of the lognormal at $\lambda = 0$, apart from the fact that it is quite hard to compute. Consider [62] or [8].

6.3 Laplace transform and fractional moments

Recall that problem (6.1) consists of finding a density $f_X(x)$ of $[0, \infty)$ such that

$$\phi(a) = E[e^{-aX}] = \int_0^\infty e^{ax} f_X(x) dx.$$

The idea of regarding the values of the Laplace transform of X as fractional moments of $Y = e^{-X}$ is not new. See for example the lecture notes in [6]. Notice that the change of variable $y = e^{-x}$ maps $[0, \infty)$ bijectively onto $(0, 1]$. The Jacobian $1/y$ of the transformation will be made part of the new density as $f_Y(y) = f_X(-\ln y)/y$, and (6.1) becomes a fractional moment problem consisting of:

Find a density $f_Y(y)$ on $[0, 1]$ such that $\quad \phi(a_i) = \int_0^1 y^{a_i} f_Y(y) dy; \quad i = 1, \ldots, K.$ (6.7)

Once such f_Y is determined, the desired density is $f_X(x) = e^{-x} f_Y(e^{-x})$.

In Chapter 4 we listed some typical problems in which the unifying theme was that $\phi(\alpha)$ could be calculated indirectly, be it analytically, or numerically by solving some equation or by simulation. Except in the case in which $\phi(\alpha)$ can be determined as a function of $\alpha > 0$, in all other cases that we deal with, it will be known only at a finite collection $\alpha_i, i = 1, \ldots, K$ of values of the transform parameter. That is, we only have at hand a few values of the transform (or a few fractional moments), and to make the problem close to real applications, the values may be known up to a range or have error built into them.

The method that we propose to deal with (6.7) has the flexibility of dealing with the two issues just mentioned. The possibility of using the method of maximum entropy in this framework was explored in [91].

6.4 Unique determination of a density from its fractional moments

Since an analytic function defined in a domain of the complex plane is determined by its values on a sequence of points having an accumulation point in the domain, it is natural to ask whether there is a connection between this fact and the Laplace inversion problem. The answer is yes, and it comes via the connection between the Laplace inversion problem and the fractional moment problem.

Before getting into that, let us recall the statement of the determinateness of the fractional moment problems in [64]:

Theorem 6.3. *Let F_Y be the distribution of a positive random variable Y. Let α_n be a sequence of positive and distinct numbers in $(0, A)$ for some A, satisfying $\lim_{n \to \infty} \alpha_n = \alpha_0 < A$. If $\mathrm{E}[Y^A] < \infty$, the sequence of moments $\mathrm{E}[Y^{\alpha_n}]$ characterizes F_Y.*

Theorem 6.4. *Let F_Y be the distribution of a random variable Y taking values in $[0, 1]$. Let α_n be a sequence of positive and distinct numbers satisfying $\lim_{n \to \infty} \alpha_n = 0$ and $\sum_{n \geq 1} \alpha_n = \infty$. Then the sequence of moments $\mathrm{E}[Y^{\alpha_n}]$ characterizes F_Y.*

So, the issue of whether values of the Laplace transform along the real axis only suffice to uniquely determine a probability density is settled. The question is now how to go about determining the density.

6.5 The Laplace transform of compound random variables

Since we are going to be using Laplace transform to determine loss distributions, it will be convenient to gather in one place some basic results and examples about the first step in the process, namely that of the computation of the Laplace transform of a compound random variable. We shall also complete here some pending computations mentioned in Section 5.3. For the sake of ease of presentation, we shall repeat some of the material introduced in Chapters 3 and 4.

6.5.1 The use of generating functions and Laplace transforms

Let us begin with the following definition:

Definition 6.2. When N is a random variable that assumes only positive integer values, with probabilities $p_k = P(N = k)$, its generating function and Laplace transform were defined by

$$G_N(z) = E[z^N] = \sum_{n=0}^{\infty} z^n p_n, \quad \text{with } z \in \mathbb{C}, \ |z| < 1,$$

and for $z = e^{-\alpha}$ with $\alpha > 0$, the Laplace transform of N is defined by

$$\phi_N(\alpha) = G_N(e^{-\alpha}) = \sum_{n=0}^{\infty} e^{-n\alpha} p_n.$$

We also mentioned that the functions $G_N(z)$ and/or $\phi_N(\alpha)$ have the following interesting properties:

Lemma 6.2. (1) *If we know $G_N(z)$ or $\phi(\alpha)$, the p_n are obtained from*

$$p_n = \frac{1}{n!} \frac{d^n}{dz^n} G_N(z) \bigg|_{z=0}.$$

(2) *If $N_1, N_2, \ldots$ is a finite or infinite collection of independent random variables such that $\sum_k N_k$ is well defined, then*

$$G_{\sum_k N_k}(z) = \prod_{k \geq 1} G_{N_k}(z),$$

$$\phi_{\sum_k N_k}(\alpha) = \prod_{k \geq 1} \phi_{N_k}(\alpha).$$

Definition 6.3. If X is a positive random variable with cumulative distribution function $F_X(x) = P(X \leq x)$, its Laplace transform is defined by

$$\phi_X(\alpha) = E[e^{-\alpha X}] = \int_0^{\infty} e^{-\alpha x} dF_X(x), \quad \text{with } \alpha > 0,$$

and when X has a density f_X the former becomes

$$\phi_X(\alpha) = E[e^{-\alpha X}] = \int_0^{\infty} e^{-\alpha x} f_X(x) dx.$$

Clearly when the values of X are integers, we reobtain the previous definition. The properties of $\phi_X(\alpha)$ are similar to those of the generating function. Consider the following statement to record this fact:

Lemma 6.3. *With the notations introduced above*
(1) $\phi_X(\alpha)$ *determines* $F_X(x)$.
(2) *The function* $\phi_X(\alpha)$ *is continuously differentiable on* $(0, \infty)$ *and*

$$E[X^n] = (-1)^n \frac{d^n}{d\alpha^n} \phi_X(\alpha) \bigg|_{\alpha=0}.$$

(3) *If* $X_1, X_2, \ldots$ *is a finite or infinite collection of positive random variables, such that* $\sum_k X_k$ *is well defined,*

$$\phi_{\sum_k X_k}(\alpha) = \prod_{k \geq 1} \phi_{X_k}(\alpha).$$

To complete the preamble we recall the following result:

Lemma 6.4. *Let N be a positive integer valued random variable and $\{X_k : k \geq 1\}$ a collection of positive, IID random variables, independent of N, and let $S = \sum_{k=1}^{N} X_k$. Then*

$$\psi_S(\alpha) = E[e^{-\alpha S}] = \sum_{k=0}^{\infty} E[e^{-\alpha S_k}] p_k = \sum_{k=1}^{\infty} (E[e^{-\alpha X_1}])^k p_k = G_N(E[e^{-\alpha X_1}]). \quad (6.8)$$

The proof is simple and uses the independence properties and the tower property of conditional expectations:

$$E[e^{-\alpha S}] = E[E[e^{-\alpha S} \mid N]] = \sum_{k=1}^{\infty} E[e^{-\alpha S_k}] P(N = k).$$

To obtain the results necessary for the gamma approximation, note that after taking natural logarithms in the previous identity, the successive derivatives up to order 3 of the left-hand side yield the centered moments of S. To compute the derivative of the logarithm of the right-hand side, just apply the chain rule and keep in mind that $G_N(\phi(\alpha)) = E[\phi(\alpha)^N]$.

In order to prove the following identities, we need some work to justify that differentiation can be exchanged with integration at all stages. That is systematic but somewhat tedious and we shall skip it. It takes a computation to verify that

$$-\frac{d\phi_S(\alpha)}{d\alpha} = \frac{E[N\phi_X(\alpha)^N]}{\phi_S(\alpha)} \left(-\frac{\phi_X'(\alpha)}{\phi_X(\alpha)}\right).$$

This identity computed at $\alpha = 0$ yields $E[S] = E[N]E[X]$. To compute

$$(-1)^2 \frac{d^2 \phi_S(\alpha)}{d\alpha^2}$$

note that

$$-\frac{d}{d\alpha}\frac{E[N\phi(\alpha)^N]}{\phi_S(\alpha)} = \frac{E[N^2\phi_X(\alpha)]}{\phi_S(\alpha)\phi_X(\alpha)}\left(-\frac{\phi'_X(\alpha)}{\phi_X(\alpha)}\right) - \frac{E[N\phi(\alpha)^N]}{\phi_S(\alpha)}\left(-\frac{\phi'_X(\alpha)}{\phi_X(\alpha)}\right),$$

which, multiplied by $-\phi'_X(\alpha)/\phi_X(\alpha)$ and evaluated as $\alpha = 0$, yields $V(N)E[X^2]$. The derivative of the remaining term is

$$-\frac{d}{d\alpha}\frac{\phi'_X(\alpha)}{\phi_X(\alpha)},$$

where we already know that at $\alpha = 0$ it equals the variance $V(X)$. This, when multiplied by $E[N\phi_X(\alpha)^N]/\phi_S(\alpha)$ evaluated at $\alpha = 0$, yields $E[N]V(X)$. Putting all of this together we obtain

$$E[(S - E[S])^2] = V(N)E[X^2] + E[N]V(X).$$

We leave the verification of the case of the third derivative for the reader. Just differentiate all terms of the second derivative, use the chain rule and evaluate at $\alpha = 0$.

6.5.2 Examples

Using the examples considered in Chapters 2 and 3, we can compute the Laplace transform of some compound random variables. We already know that the Laplace transform of a compound variable is the composition of the moment generating function of the frequency variable with the Laplace transform of the individual losses. The result of that composition provides the input for the maxentropic methodology to compute the cumulative distribution function F_S or the probability density f_S of the compound losses.

Preliminary remark
At this point we mention again our aim is determine the probability density of aggregate losses given that loss occurs. According to Lemma 5.1, in the examples to obtain the Laplace transform of that conditioned random variable we should consider

$$\phi_S(\alpha) = \frac{\psi_S(\alpha) - P(N = 0)}{1 - P(N = 0)}. \tag{6.9}$$

That is why in the examples presented below we consider $P(N = k)/(1 - P(N = 0))$.

Example: Poisson frequency and gamma individual losses
Suppose that the frequency is modeled by a Poisson(λ) and the individual loses are modeled by a $\Gamma(a, b)$. In this case the Laplace transform of the compound random variable is

$$\psi_S(\alpha) = e^{-(1-(\frac{b}{b+\alpha})^a)}. \tag{6.10}$$

As we mentioned, this is an example that is easy to study, because the convolution of a gamma density with itself is again a gamma density. This example was already mentioned in Section 5.3.1. There we pointed out that the probability density, conditioned on the occurrence of losses, is given by

$$f_S(x) = \sum_{k\geq 1} p_k^0 \left(\frac{b^k x^{(ak-1)}}{\Gamma(ak)} \right) e^{-bx}$$

and we saw there that the Laplace transform of this density is given by

$$\frac{\psi(\alpha) - e^{-\lambda}}{1 - e^{-\lambda}}.$$

This is a series representation of the density of aggregate loses, which can be used to compare with the result of the maxentropic procedure.

Example: Poisson frequency and binary individual loses

This is a very simple example that illustrates the use of the composition law of generating functions. Suppose that the number of losses during a time period is random, with a Poisson(λ) frequency, and when a loss event occurs, either we lose 0 with probability p or we lose 1 with probability $1-p$. Now $G_N(z) = \exp(-\lambda(1-z))$ and $\phi_X(\alpha) = p + (1-p)e^{-\alpha}$. In this case it is easy to see that

$$\psi_S(\alpha) = \exp(-(1-p)\lambda(1-e^{-\alpha})),$$

which is the Laplace transform of an integer valued random variable of Poisson type with frequency $(1-p)\lambda$.

Example: binomial frequency and gamma individual losses

Suppose this time that there is a finite number of losses, occurring according to a $B(n,p)$ distribution and that individual losses are as above. This time the Laplace transform of the compound variable looks like

$$\psi_S(\alpha) = \left(p\left(\frac{b}{b+\alpha}\right)^a + (1-p) \right)^n. \tag{6.11}$$

Now the series expansion of the compound density has a finite number of terms, and the analysis and comparisons with the output of the maxentropic procedure are even easier to carry out. It is not hard to see that

$$f_S(x) = \sum_1^n \frac{p^k(1-p)^{(n-k)}}{1-(1-p)^n} \binom{n}{k} \left(\frac{b^k x^{(ak-1)}}{\Gamma(ak)} \right) e^{-bx}.$$

Example: binomial frequency and uniform losses
In the simple model of credit risk, in which there are n independent borrowers, each having a probability p of defaulting, in which case the bank will sustain a loss uniformly distributed on $[0,1]$, the Laplace transform of the loss sustained by the bank is given by

$$\psi_S(\alpha) = \left(p\left(\frac{1-e^{-\alpha}}{\alpha}\right) + (1-p)\right)^n = \sum_{k=0}^{n} p^k (1-p)^{(n-k)} \left(\frac{1-e^{-\alpha}}{\alpha}\right)^k. \tag{6.12}$$

Here we made use of the fact that the Laplace transform of a $U[0,1]$ random variable is $\phi_X(\alpha) = (1-e^{-\alpha})/\alpha$. Again, there is not much numerical nor practical difficulty here. To invert this Laplace transform note that

$$\left(\frac{1-e^{-\alpha}}{\alpha}\right)^k = \frac{1}{\alpha^k} \sum_{j=0}^{k} (-1)^j \binom{k}{j} e^{-j\alpha}$$

and note as well that

$$\frac{1}{\alpha^k} e^{-j\alpha} = \int_0^\infty \frac{(x-j)_+^{k-1}}{(k-1)!} e^{-x\alpha} dx,$$

where we use the notation $(a)_+$ to denote the positive part of the real number a. Putting these two remarks together, it follows that the sum of k independent random variables uniformly distributed on $[0,1]$ has a density (on $[0,k]$) given by

$$f_k(x) = \frac{1}{(k-1)!} \sum_{0 \leq j \leq x} (-1)^j \binom{k}{j} (x-j)^{(k-1)}$$

for $0 < x < k$ and it equals 0 otherwise. A finite weighted sum of such polynomials presents no difficulty and can be used for comparison purposes. As there is a probability $(1-p)^n$ of no default occurring, the conditional density of losses, given that losses occur, according to Lemma 5.1 is given by

$$f_S(x) = \sum_{k=1}^{n} \frac{p^k (1-p)^{(n-k)}}{1-(1-p)^n} \binom{n}{k} f_k(x),$$

where the $f_k(x)$ are given a few lines above.

6.6 Numerical determination of the Laplace transform

To repeat ourselves once more, the motivation for this section is the fact that we may not be able to compute a Laplace transform, even if we happen to know the distribution function of a random variable, like in the case of the lognormal density, or we may only know a differential equation satisfied by the density, or by the Laplace transform itself, or as in the case to be considered, we may just have a sample of the random variable itself. The sample may come from actual measurements or be the result of a simulation process.

Let us now examine how to determine the value of the Laplace transform of a positive random variable from real or simulated data. So, let us suppose that S is a positive random variable. Suppose that we have a sample $\{s_1, \ldots, s_M\}$ of size M of S, which is the result of M independent observations of S. The estimated Laplace transform is defined by

$$\widehat{\psi}(\alpha) = \frac{1}{M} \sum_{j=1}^{M} e^{-\alpha s_j}. \tag{6.13}$$

There is nothing special about this definition. It just relies on the law of large numbers for its verification, and on the central limit theorem for a measure of the quality of the approximation, or of the error in the approximation, if you prefer.

With a eye on what comes below, that is the determination of the probability density of the losses, notice that if S models losses in a given time period, and if there might periods in which no losses occur, we have to weed out this part of the above computation. For that, let M_0 be the number of sample points at which $s_j = 0$, and let M_+ be the number of nonzero sample points. Certainly, $M = M_0 + M_+$. Notice now that

$$\widehat{\psi}(\alpha) = \frac{1}{M} \sum_{j=1}^{M} e^{-\alpha s_j} = \frac{M_0}{M} + \frac{M_+}{M} \frac{1}{M_+} \sum_{s_j > 0} e^{-\alpha s_j},$$

which can be rewritten as follows:

$$\widehat{\phi}(\alpha) = \frac{\widehat{\psi}(\alpha) - \widehat{P}(N=0)}{1 - \widehat{P}(N=0)} = \frac{1}{M_+} \sum_{s_j > 0} e^{-\alpha s_j}. \tag{6.14}$$

Here we have set $\widehat{P}(N = 0) = M_0/M$. This is the empirical analogue of Lemma 5.1, and this will be the input for the fractional moment problem in the numerical procedures to determine the density of losses from actual data. In Chapter 11 we shall take up the problem of the influence of the sample dependence on the reconstructed density.

6.6.1 Recovering a density on [0, 1] from its integer moments

We have already mentioned that a Laplace inversion problem can be mapped onto a fractional moment problem for a density on [0,1]. This will be useful when we only know the Laplace transform at a few, nonintegral values of the variable.

We have also related the problem of the inversion of the Laplace transform when the values of the moments $E[X^n]$ are available to us. Here we describe a technique based on the knowledge of the Laplace transform at integral values of the parameter, which can be invoked to invert it. It also relies on the mapping of the Laplace transform X at $\alpha = n$ to the moment of order n of $Y = e^{-X}$.

Let us consider a continuous probability density $g(y)$ on $[0, 1]$. Notice that it can be uniformly approximated by a step function

$$g_N(y) = \sum_{i=1}^{N} g_N(i) I_{((i-1)/N, i/N]}(y).$$

The method we examine consists of relating the values of $g_N(i)$ to the integral moments of Y, and it is made explicit below. But first, let us introduce some notations. Write

$$\mu_n = \int_0^1 y^n g(y) dy$$

for the integer moments of Y, then the result that we want is contained in:

Theorem 6.5. *Let g be a continuous density on $[0, 1]$ with moments μ_n, for $n \geq 1$. Then for $0 < \delta < 1$, and, given N we set*

$$g_N(i) = (N+1) \binom{N}{i} \sum_{k=0}^{N-i} \binom{N-i}{k} (-1)^k \mu_{k-i}$$

and the function g_N is defined as above, then

$$\|g - g_N\| \leq \Delta(g, \delta) + \frac{2\|g\|}{\delta(M+2)},$$

where $\Delta(g, \delta) := \sup\{|g(y) - g(y')| \mid |y - y'| < \delta\}$.

Comment. We use $\|h\|$ for the uniform or L_∞ norm of h. This result is proved in [72]. See as well [73] for the case in which the change of variables $f(x) = e^{-x} g(e^{-x})$ is undone as part of the process on inversion.

6.6.2 Computation of (6.5) by Fourier summation

Here we present an approximate method for evaluating (6.5). This is useful when the Laplace transform of a density is known to be, or can be extended to, an analytic function on the right half complex plane, and all of its singularities are poles on the left half complex plane.

6.6 Numerical determination of the Laplace transform

Note that a simple change of sign in the integration variable allows us to write

$$\int_{-\infty}^{+\infty} e^{x(a+is)}\mu(a+is)ds = \int_{-\infty}^{0} e^{x(a+is)}\mu(a+is)ds + \int_{0}^{+\infty} e^{x(a+is)}\mu(a+is)ds$$

$$= 2e^{xa}\int_{0}^{+\infty}\mathbb{R}(e^{ixs}\mu(a+is))ds.$$

Therefore, the classical inversion formula becomes

$$f(x) = \frac{1}{\pi}e^{xa}\int_{0}^{+\infty}\mathbb{R}(e^{ixs}\mu(a+is))ds.$$

This is the starting point for Crump's approximation of the contour integral by a Fourier series; see [25]. An application of the trapezoidal rule, with steps of size π/T transforms the previous integral into

$$f(x) = \frac{e^{xa}}{T}\left(\frac{\mu(a)}{2} + \sum_{k=1}^{\infty}\mathbb{R}\left\{\mu\left(a+i\frac{k\pi}{T}\right)\exp\left(i\frac{k\pi}{T}\right)\right\}\right). \qquad (6.15)$$

Concluding Remark. The methods described in the last two sections, useful as they are, depend on the knowledge of $\mu(n)$ (or $\psi(a)$) at quite a large number of points. In the next chapter, we shall compare the reconstructions based on the these two methods with the reconstructions obtained using the maxentropic technique, which makes use of a very small number of data points.

7 The standard maximum entropy method

We begin this chapter with some historical remarks about the concept of entropy of a probability distribution. We decided to do this because now and then a reviewer of a paper, or a colleague who vaguely remembers his high school chemistry classes, has asked us what the problems addressed here have to do with problems in physical chemistry. The answer is nothing of course, even though the origin of the name comes from the natural sciences. We find these comments pertinent and we hope that they are welcome by colleagues in the banking and the insurance industries.

The concept, or should we better say, the name 'entropy' has been used (and misused) in many different contexts: in art, economics, politics, besides being an important concept in information theory (the transmission of information), statistics and in natural sciences. And there is even an open access journal, *Entropy*, devoted to all sort of applications of the concept and of the method of maximum entropy. We mention two interesting expository books, [15] and [9], in which the concept of entropy plays a central role.

From the historical point of view, the concept of entropy originates in physical chemistry, and the notion of entropy as a function of some probability distribution comes from Boltzmann around 1875. He was studying the dynamics of the approach to equilibrium of a gas. There he proposed understanding the macroscopic behavior of a system with a very large number of particles through the study of a density $f(t, x, v)$, which describes the number of particles in a volume $dxdv \in V \times \mathbb{R}^3$ of the 'configuration space' (position and velocities of the system). We denote by V the volume that the particles occupy. Boltzmann proposed a 'dynamics' for $f(t, x, v)$, that is an equation that explains how it changes due to the action of internal and external forces on the particles and due to the collisions of the particles among themselves. To understand the approach to equilibrium, Boltzmann noticed that the function

$$S(f_t) = - \int_{V \times \mathbb{R}^3} f(t, x, v) \ln f(t, x, v) dx dv$$

behaved as a Lyapunov function. That is, he noticed that under the proposed dynamics, $S(f_t)$ increased to the static (equilibrium) solution of the equation.

Later on, at the turn of the 20th century, came the work by Gibbs on the foundation of statistical mechanics. He used the same functional form, but he considered the probability densities to be defined in 'phase space', which is $V^N \times \mathbb{R}^{3N}$. Again, the state of equilibrium of the system corresponds to the density that maximizes the entropy function.

The same concept appeared in the late 1940s in the work of Shannon [87], who was designing efficient ways to deal with the noise in the transmission of messages. During the late 1950s the same concept appeared in the field of statistics; see [61]. But it seems that an implementation of Gibbs' ideas by Jaynes [57] led to the variational method in the way that it is used nowadays. It is a procedure to solve the *generalized moment problem*, which consists of determining a probability distribution or a probability density from the knowledge of the expected values of a few random variables. The volumes [24] and [58] collect much of the work done on applications of the concept of entropy in information theory, statistics, and physics.

Besides that, the concept of entropy appears in a curious and perhaps unexpected way in large deviations theory; see [33].

7.1 The generalized moment problem

Let us state the problem that we want to solve in its simplest and direct form. In probabilistic modeling one starts with a collection of observed quantities $X_i: i = 1, \ldots, M$ that assume values in subsets $I_i \subset \mathbb{R}, i = 1, \ldots, M$. Usually, such subsets are (bounded or unbounded) intervals, and with them we form the sample space $\Omega = \prod_{i=1}^{M} I_i$, along with a σ-algebra $\mathcal{F}$ of subsets of Ω. The questions that one asks about experiments involving the random quantities are of the following type: What is the fraction of observations in which $\{X_i \in J_i\}$, where $J_i \in \mathcal{F}$? To answer these questions, one needs to have a probability on $(\Omega, \mathcal{F})$, and it is at this point where the maximum entropy methodology comes in.

The way this goes is as follows. Instead of measuring $P(X_i \in J_i)$ for all possible J_i we set up a reduced problem to solve, but which provides us with a good answer. We are thus led to a problem that can be stated as: Determine a probability P on a measure space $(\Omega, \mathcal{F})$, from the knowledge of the expected values of a few random variables, that is from

$$E_P[h_i(\boldsymbol{X})] = \mu_i, \quad \text{for } i = 1, \ldots, K, \tag{7.1}$$

where the functions $h_i: \Omega \to \mathbb{R}; i = 1, \ldots, K$ are chosen by the modeler to be as informative about $\boldsymbol{X}$ as possible. This problem is called the generalized moment problem, and it is in solving that problem that the method of maximum entropy comes in.

In order to make our life simpler, and to not start in a vacuum, we suppose that the probability P is absolutely continuous with respect to a measure m defined on $(\Omega, \mathcal{F})$, which we call the *reference measure*. That is, we constrain our quest to densities f such that $dP = fdm$. In many problems, when Ω is an open or closed subset of $\mathbb{R}^d$, m is chosen to be the Lebesgue measure, (the usual $d\boldsymbol{x}$), or when Ω is a countable set, it is just the counting measure. Sometimes we may find it convenient to let m be a probability on $(\Omega, \mathcal{F})$ in which some prior information about the law P is incorporated.

When we limit our search to probabilities having a density f with respect to m, the statement of the problem becomes: Find a density f such that

$$E_P[h_i(\boldsymbol{X})] = \int_\Omega h_i(\boldsymbol{X}(\omega))f(\omega)m(d\omega) = \mu_i, \quad \text{for } i = 1, \ldots, K. \tag{7.2}$$

It is rather simple to verify the following lemma:

Lemma 7.1. *With the notations introduced above we have:*
(1) *The set of probabilities satisfying (7.1) is convex in the class of probabilities that are absolutely continuous with respect to the measure m.*
(2) *The set of densities satisfying (7.2) are convex sets in the class of all $L_1(m)$ integrable functions.*

Here we use the usual notation $L_1(m)$ to refer to the class of all $\mathcal{F}$ measurable functions h on Ω such that $\int_\Omega |h| dm < \infty$. But we mention at this point that the class of densities (or the class of nonnegative integrable functions for that matter) is not an open set in the topology defined by the $L_1(m)$ norm. To see this, just add a small (in norm) nonpositive perturbation to a positive function. The perturbed function stays inside any small ball in $L_1(m)$, but it is not positive. This is an important technical detail overlooked in many applications of the method.

The basic intuition behind the method of maximum entropy is that in order to determine a density satisfying (7.2), it suffices to define a concave function of the class of all densities and prove that it achieves a maximum in the class satisfying (7.2). Certainly there are many possible functions to propose, and it so happens that the entropy function has proven to be a good choice. In order to solve the generalized moment problem, we need to properly define the entropy functions and examine their properties, and then use them to solve the problem.

7.2 The entropy function

We shall suppose that all measures and probabilities that we talk about are defined on a measurable space $(\Omega, \mathcal{F})$. We begin with a general definition and then consider variations on the theme.

Definition 7.1. The entropy of a probability P relative to a measure m is defined by

$$S_m(P) = \begin{cases} -E_P[\ln(\frac{dP}{dm})] = -\int \frac{dP}{dm} \ln(\frac{dP}{dm}) dm, & \text{if } P \ll m, \text{ and the integral is finite} \\ -\infty, & \text{otherwise.} \end{cases} \quad (7.3)$$

The standing convention here is that $0 \ln 0 = 0$. Also, if m is a counting measure concentrated on a set $\{\omega_1, \omega_2, \dots\}$, and if $P(\omega_i) \neq 0$, then the definition becomes

$$S_m(P) = -\sum_{i \geq 1} P(\omega_i) \ln P(\omega_i).$$

But, as we are going to be dealing with problem (7.2), we shall be more concerned with the following version of (7.3). Let $dP = fdm$. The entropy of the density f is defined as follows:

$$S_m(f) = \begin{cases} -\int f \ln(f) dm, & \text{if the integral is finite} \\ -\infty, & \text{otherwise.} \end{cases} \tag{7.4}$$

We call it the *entropy of f* without further reference to m when it stays fixed. For some applications it is convenient to consider m as a probability measure; let us denote it by Q and say that its choice is up to the modeler. In this case (7.3) is replaced by

$$S_Q(P) = \begin{cases} -E_P[\ln(\frac{dP}{dQ})] = -\int \frac{dP}{dQ} \ln(\frac{dP}{dQ}) dQ, & \text{if } P \ll Q, \text{ and the integral is finite} \\ -\infty, & \text{otherwise.} \end{cases} \tag{7.5}$$

In this case one usually writes $S(P, Q)$ and calls it the *entropy of P with respect to Q*. As last variation on the theme, consider the case in which $P \ll Q$ and both P and Q have densities with respect to a reference measure m, that is, $dP/dm = f$ and $dQ/dm = g$. In this case $S(P, Q)$ becomes

$$S_m(P, Q) = \begin{cases} -E_P[\ln(\frac{dP}{dQ})] = -\int \frac{f}{g} \ln(\frac{f}{g}) g \, dm, & \text{if the integral is finite} \\ -\infty, & \text{otherwise.} \end{cases} \tag{7.6}$$

Now we call $S_m(f, g)$ the *entropy of f with respect to g*. At this point we mention that beginning with the work of Kullback, in statistics and information theory it is customary to work with $K(f, g) = -S_m(f, g)$, which then is convex in f.

Theorem 7.1. *With the notations introduced above, we have:*
(1) *The function $f \to S_m(f, g)$ is strictly concave.*
(2) $S_m(f, g) \leq 0$, *and* $S_m(f, g) = 0$ *if and only if* $f = g$ *almost everywhere with respect to m.*
(3) *When $S_m(f, g)$ is finite, we have (Kullback's inequality)*

$$\frac{1}{4} \|f - g\|^2 \leq -S_m(f, g) = K(f, g).$$

This result is proved in several places, for example in [61], [13] or [24]. A sketch of the proof goes as follows: First note that the function $t \to -t \ln t$ is strictly concave. This property is preserved when t is replaced by a function and integrated. The second property is an easy consequence of Jensen's inequality. The third assertion is a guided exercise in Chapter III of [61].

Let us now introduce some more notation. With the notations introduced for (7.1) or (7.2),

$$\mathcal{P} = \{P \ll m \mid (7.1) \text{ or } (7.2) \text{ holds}\}. \tag{7.7}$$

Consider now the following exponential family of densities (called the Hellinger arc of X):

$$\rho_\lambda = \frac{e^{-\langle\lambda,h(X)\rangle}}{Z(\lambda)} \tag{7.8}$$

$$Z(\lambda) = \int_\Omega e^{-\langle\lambda,h(X)\rangle}\, dm. \tag{7.9}$$

The normalization factor is defined on

$$\mathcal{D}(m) = \{\lambda \in \mathbb{R}^K \mid Z(\lambda) < \infty\}. \tag{7.10}$$

The first important property of $Z(\lambda)$ is contained in:

Proposition 7.1. *With the notations just introduced we have:*
(1) *The set $\mathcal{D}(m)$ is convex and the function $\lambda \to \ln Z(\lambda)$ is convex on $\mathcal{D}(m)$.*
(2) *For $f \in \mathcal{P}$ and $\lambda \in \mathcal{D}(m)$,*

$$S_m(f) \leq \Sigma(\lambda,\boldsymbol{\mu}) \equiv \ln Z(\lambda) + \langle\lambda,\boldsymbol{\mu}\rangle.$$

The proof of the first assertion emerges from Hölder's inequality, and the proof of the second consists of an invocation of Theorem 7.1 applied to f and g replaced by ρ_λ, and the fact that f satisfies (7.2). Note that if $\rho_\lambda \in \mathcal{P}$, then $\Sigma(\lambda,\boldsymbol{\mu}) = S_m(\rho_\lambda)$.

7.3 The maximum entropy method

We have already mentioned the intrinsic difficulty in doing calculus on the space of densities, in particular if we tried to find the maximum of a convex function defined on a convex subset of the class of densities. The difficulty lies in the fact that an arbitrarily small (in the $L_1(m)$-norm) variation $f \to f + \delta f$ takes the function out of the class of densities. This makes the application of the concepts of calculus hard to apply in this infinite dimensional setup.

But there is a way out of this difficulty. Notice that the inequality in Proposition 7.1 asserts that the entropy $S_m(f)$ of any density f satisfying the constraints is bounded above by the entropy $S_m(\rho_\lambda)$. The problem is that we do not know whether ρ_λ satisfies (7.2). If we were able to prove that for some λ^*, the density ρ_λ satisfies (7.2) or is in $\mathcal{P}$, then we will be done. Then the natural thing to do is to minimize $\Sigma(\lambda,\boldsymbol{\mu})$ and verify that for the λ^* at which the minimum is reached, the density $\rho_{\lambda^*} \in \mathcal{P}$. So, let us study some of its properties. We begin with the following assumption:

Assumption A. *Suppose that the convex set $\mathcal{D}(m)$ has a nonempty interior in $\mathbb{R}^K$.*

Thus it is clear that

$$\int_\Omega e^{-\langle\lambda,h(X)\rangle}\, dm \tag{7.11}$$

is continuously differentiable in $\mathcal{D}(m)$ as many times as we need, but all that we need for the applications that come below is:

Proposition 7.2. *As function of* (λ, μ)*, the function* $\Sigma(\lambda, \mu)$ *has two continuous derivatives in* $\mathcal{D}(m)$*, and*

$$-\frac{\partial}{\partial \lambda_i} \ln Z(\lambda) = \int_\Omega h_i(X) \frac{e^{-\langle \lambda, h(X)\rangle}}{Z(\lambda)} dm.$$

$$\frac{\partial^2}{\partial \lambda_i \partial \lambda_j} \ln Z(\lambda) = \int_\Omega h_i(X) h_j(X) \frac{e^{-\langle \lambda, h(X)\rangle}}{Z(\lambda)} dm$$
$$- \int_\Omega h_i(X) \frac{e^{-\langle \lambda, h(X)\rangle}}{Z(\lambda)} dm \int_\Omega h_j(X) \frac{e^{-\langle \lambda, h(X)\rangle}}{Z(\lambda)} dm. \tag{7.12}$$

The dependence of $\Sigma(\lambda, \mu)$ on μ will come into focus below, when we study problems with data in intervals. The result that we need for the time being is the following:

Theorem 7.2. *With the notations introduced above and under Assumption A, suppose that the minimum of* $\Sigma(\lambda, \mu)$ *as function of* λ *is achieved at* λ^* *in the interior of* $\mathcal{D}(m)$*, then*

$$-\nabla_\lambda \ln Z(\lambda^*) = \mu. \tag{7.13}$$

Then the density

$$\rho_{\lambda^*} = \frac{e^{-\langle \lambda^*, h(X)\rangle}}{Z(\lambda^*)} \tag{7.14}$$

solves the entropy maximization problem, and furthermore

$$S(\rho_{\lambda^*}) = \Sigma(\lambda^*, \mu) = \ln Z(\lambda^*) + \langle \lambda^*, \mu \rangle. \tag{7.15}$$

Proof. Notice that if λ^* is the minimizer of $\Sigma(\lambda, \mu)$ for fixed μ, then, invoking the first identity in (7.12), we have

$$\nabla_\lambda \Sigma(\lambda^*, \mu) = 0 \quad \Leftrightarrow \quad -\nabla_\lambda \ln Z(\lambda^*) = \mu.$$

In particular, this asserts that ρ_λ^* is a density that solves (7.2). The third assertion is obtained by a simple computation. Not only that, according to the comment right after the proof of Proposition 7.1, we conclude that at ρ_λ^* the entropy attains its maximum possible value. □

Comment. In the statement of the theorem we considered the nicest possible case. We can be a bit more general and suppose that the optimal λ^* is achieved at the relative interior of $\mathcal{D}(m)$, and still obtain the same result. But it may happen that for fixed μ, the function $\lambda \to \Sigma(\lambda, \mu)$ is not bounded below. In this case, which may occur when there is inconsistency of the data, the maximum entropy method breaks down. Rigorous proofs of all these assertions appear in [13]. As to the use of duality to obtain the result as we did, it seems that it was in [70] that a connection to duality theory was first mentioned.

7.4 Two concrete models

To close, we shall make explicit the choice of $(\Omega, \mathcal{F})$ for two possible cases. Let us begin by the simplest case.

7.4.1 Case 1: The fractional moment problem

This example is a particular case of the example just described. Here we shall consider $M = 1$ and $\Omega = [0, 1]$ with the σ-algebra $\mathcal{F} = \mathcal{B}([0, 1])$ and the reference measure $dm(y) = dy$, the standard Lebesgue measure on the interval. To vary the notation a bit, now (7.2) becomes: Find a density $f(y)$ on $[0, 1]$ such that

$$\int_0^1 y^{\alpha_i} f(y) dy = \mu_i; \quad i = 1, \ldots K. \tag{7.16}$$

That is, in this case $h_i(y) = y^{\alpha_i}$. We already saw that the solution of the associated maximum entropy problem is explicitly given by

$$f^*(y) = \frac{e^{-\sum_{i=1}^K \lambda_i^* y^{\alpha_i}}}{Z(\lambda^*)}, \tag{7.17}$$

where λ^* is obtained as described, and the particular aspect of $Z(\lambda)$ is now

$$Z(\lambda) = \int_0^1 e^{-\sum_{i=1}^K \lambda_i y^{\alpha_i}} dy.$$

7.4.2 Case 2: Generic case

In this case, the random vector $\boldsymbol{X}$ can be regarded as a vector variable taking values in $\mathbb{R}^K$, and we can think of $(\Omega, \mathcal{F})$ as $(\mathbb{R}^K, \mathcal{B}(\mathbb{R}^K))$. This time we suppose that the information available to the modeler is about the expected values of the X_i, $i = 1, \ldots, K$ themselves, that is $h_i(\boldsymbol{X}) = X_i$. We are not placing any restrictions on the range of the X_i, besides the knowledge of their mean values. Furthermore, we suppose that the modeler knows that the unknown probability distribution has an unknown density $\rho(x)$ with respect to a probability $dQ(x)$ on $\mathbb{R}^K$. From the maximum entropy procedure, we know that the unknown density is of the form

$$\rho_{\lambda^*}(x) = \frac{e^{-\langle \lambda^*, x \rangle}}{Z(\lambda^*)} \tag{7.18}$$

and, for $\lambda \in \mathcal{D}(Q)$, the normalization factor is given by

$$Z(\lambda) = \int e^{-\langle \lambda, x \rangle} dQ(x).$$

Notice that it is the choice of the reference probability $dQ(x)$ that makes the problem nontrivial. Actually, the solution to be obtained may depend strongly on the choice of the $dQ(x)$, and a theory of best choice of such reference measure is, as far as we know, an unstudied problem. Again, the notations and the procedure to determine the λ^* is as described in the previous section.

7.5 Numerical examples

In this section we shall consider several types of examples. First, we shall consider the problem of reconstructing a density from simulated data from a known distribution. Next we shall determine a distribution of losses with several levels of aggregation, and to finish we shall compare the output of the maxentropic procedure versus two other reconstruction methods.

In the first case, after explaining how the data is generated, we display the maxentropic reconstructions obtained by an application of the standard maximum entropy (SME) procedure described above, and compare the result obtained with the true distribution by means of several measures of discrepancy.

For the case in which the data is obtained as a result of several levels of aggregation, we shall compare with the empirical distribution for a case in which the amount of data is moderately large. And to finish, to compare with two nonmaxentropic procedures, we shall consider an example discussed in Chapter 5.

7.5.1 Density reconstruction from empirical data

We shall consider next a quite simple case, consisting of reconstructing the density of a positive random variable when we have a sample from its distribution. But more importantly, to illustrate the power of the method, we shall only use eight values of its Laplace transform determined from the data. The sample that we shall consider comes from a lognormal density. This example is chosen because neither its Laplace transform nor its characteristic function can be determined analytically, and we have to rely on numerical estimation. Moreover, and this is the beauty of the example, the reconstructed density function can be compared to the true one after the maxentropic procedure is carried out.

The data generation process

In order to test the robustness of the maxentropic procedure, we shall consider two samples $\{s_1, \ldots, s_N\}$, one of size $N = 200$ and the other of size $N = 1000$, from a lognormal density with parameters $\mu = 1$ and $\sigma = 0.1$. The fractional moments (or the Laplace transform) at $\alpha_i = 1.5/i$ for $i = 1, \ldots, 8$, is computed according to

$$\bar{\mu}_i = \frac{1}{N} \sum_{k=1}^{N} e^{-s_j \alpha_i}, \quad \alpha_i = 1.5/i; \quad i = 1, \ldots, 8. \tag{7.19}$$

The empirical moments obtained are listed in Table 7.1.

Table 7.1: Moments of S for different sample sizes.

Size	Moments of S							
	μ_1	μ_2	μ_3	μ_4	μ_5	μ_6	μ_7	μ_8
200	0.0176	0.1304	0.2562	0.3596	0.4409	0.5051	0.5568	0.5990
1000	0.0181	0.1319	0.2579	0.3612	0.4424	0.5066	0.5581	0.6002

In the next chapter we shall list the confidence intervals for these moments. They will play a role in the reconstruction procedure carried out there when the data is specified to be in an interval.

The maxentropic density

In Figure 7.1 we present the histograms of the two sets of data points. In the left panel we present the histogram of the sample of size $N = 200$ and to its right that of the sample of size $N = 1000$.

In the two panels of Figure 7.2 we display the maxentropic densities, and then in the two panels of Figure 7.3 the cumulative distribution functions for the two sample sizes.

In both figures the left-hand panel corresponds to the reconstruction obtained from $N = 200$ data points and the right-hand panel to the corresponding reconstruction obtained from $N = 1000$ data points.

In the plots of the cumulative distribution functions, the jagged line corresponds to the cumulative distribution of the empirical density of the data. This shows the amount of discrepancy between the data and the reconstructed density better than the visual comparison between the density and the histogram, which is bin dependent.

In Tables 7.2 and 7.3 shown below, we present some numerical measures of quality of reconstruction. In Table 7.2 we present the comparison of the L_1 and L_2 norms

7.5 Numerical examples — 85

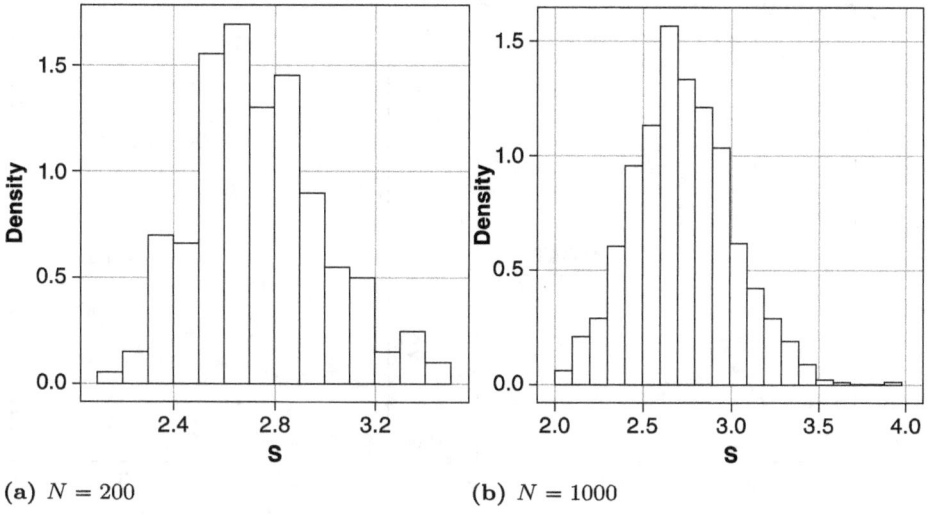

(a) $N = 200$ (b) $N = 1000$

Figure 7.1: Histograms of the two samples.

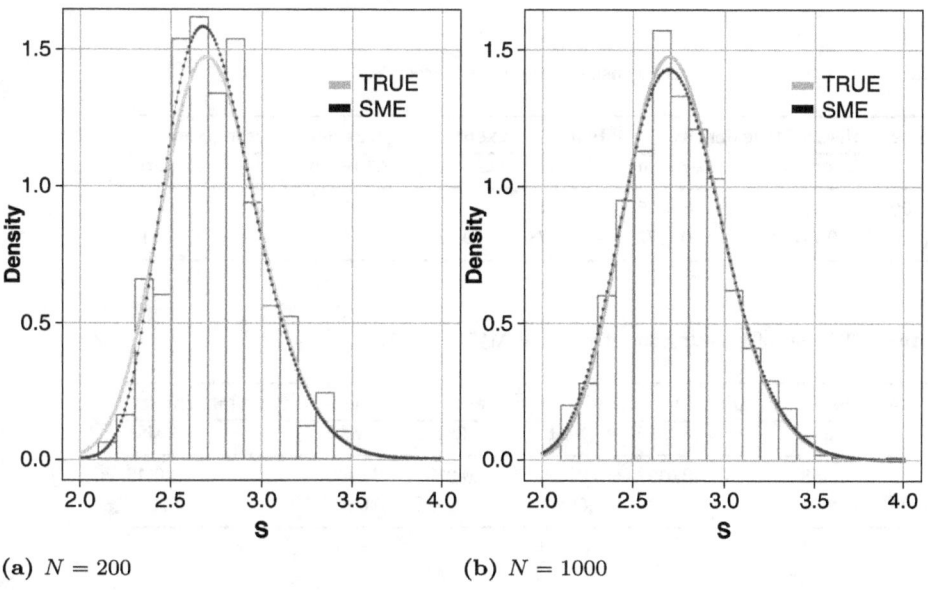

(a) $N = 200$ (b) $N = 1000$

Figure 7.2: Density distributions for different sample sizes.

between the true density and the histogram to stress the fact that this measure of discrepancy between the reconstructed density and the empirical density is bin dependent. Note that the L_1 and L_2 distances between the true and the empirical density as well as the L_1 and L_2 distances between the maxentropic and the empirical densities are quite

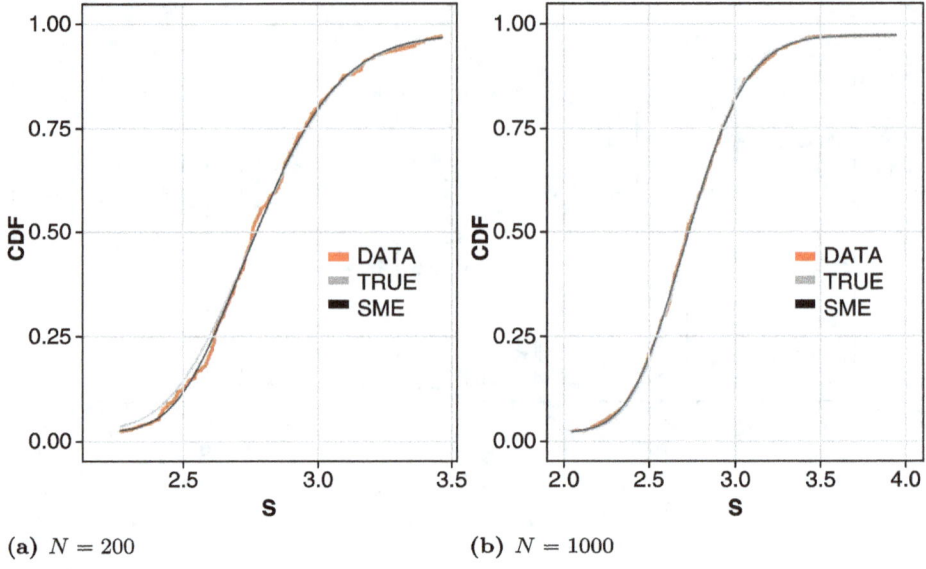

(a) $N = 200$ (b) $N = 1000$

Figure 7.3: Cumulative distribution functions for different sample sizes.

Table 7.2: Quality of reconstruction using L_1 and L_2 distances.

Size	Hist. vs. true density		Hist. vs. maxent.		True density vs. maxent.	
	L1-norm	L2-norm	L1-norm	L2-norm	L1-norm	L2-norm
200	0.1599	0.1855	0.1449	0.1727	0.0668	0.0761
1000	0.1042	0.1077	0.0973	0.1044	0.0307	0.0289

Table 7.3: Quality of reconstruction according to MAE and RMSE.

Size	Hist. vs. true density		Hist. vs. maxent.		True density vs. maxent.	
	MAE	RMSE	MAE	RMSE	MAE	RMSE
200	0.0158	0.0199	0.0089	0.0103	0.0105	0.0115
1000	0.0064	0.0076	0.0043	0.0056	0.0053	0.0060

similar. Also note that the L_1 and L_2 distances between the true and the maxentropic densities are considerably smaller. Of course, this type of comparison is not possible in 'real life' situations because we do not have the true density to begin with.

As an alternative to the L_1 and L_2 distances between densities, in Table 7.3 we present the mean absolute error (MAE) and root mean square error (RMSE) as an alternative measure of error. These two discrepancy measures are not bin dependent, and capture the closeness between the cumulative densities displayed in Figures 7.2 and 7.3.

Clearly, the results seem to be pretty good using either measure. For L_1 and L_2 norms, the reconstruction is close to the true density. This result seems to be better because we are directly comparing two densities instead of the histogram. For the MAE and RMSE measures the reconstruction is closer to the simulated data. It is important to clarify that for this calculations we obtain the cumulative distribution for each available data point, not just the centers of the bins as in the case of L_1 and L_2 norms (see Figure 7.3).

To finish, we mention a complementary result. The minimization of the dual entropy $\Sigma(\lambda, \mu)$ is carried out with a step reducing Newton gradient method available in the R library under the name *B&B*-package. The norm of the gradient of the dual entropy, which measures the degree with which the maxentropic density reproduces the data, was of the order of 10^{-7} for the sample of size $N = 200$ and 10^{-8} for the sample of size $N = 1000$.

7.5.2 Reconstruction of densities at several levels of data aggregation

In this section we address the problem of numerically determining the density of a positive random variable, which results from an underlying multistage aggregation process. Think of a bank that during a given period of time is aggregating losses of risks of different types into risk in a line of business, and then aggregating the risk of the different business lines into one total loss for the time period. Or think of an insurance company that aggregates liability payments produced by different types of claims during a given period. In these cases, depending on how it was collected, the empirical data may or may not come with the dependencies among the different losses built in.

From the point of view of evaluating the capabilities of the method of maximum entropy, the comments in the previous paragraph have to be taken into account to be sure that the right input is fed into the maxentropic reconstruction procedure.

The aggregation process that we consider has three levels. We might think of the total loss as aggregations of losses in several business lines, each of which is the result of aggregating losses of different type in each of them, and finally (or to begin with), the losses of each type result from aggregating individual severities according to the number of times that they occur during some preassigned lapse of time. In the context of claims to be paid by an insurance company, the total amount paid is the sum of claims paid by accidents, which may be of different types. The numerical example that comes was developed in [46].

To put it in symbols, for each $b = 1, 2 \ldots, B$ consider $h = 1, 2, \ldots, H_b$ and put

$$S = \sum_{b=1}^{B} S_b, \quad \text{with } S_b = \sum_{h=1}^{H_b} S_{b,h}, \tag{7.20}$$

where, for each pair (b, h),

$$S_{b,h} = \sum_{k=1}^{N_{b,h}} X_{b,h,k} \qquad (7.21)$$

describes the loss at the first level of aggregation. Here we suppose that all the individual losses are independent among themselves and of the numbers of losses in each time period, and those of the same (b, h, k) type are identically distributed.

Next we describe how the data is generated. Once the data for S is available we proceed as in the previous section, that is, we compute numerically the Laplace transform at eight values of the parameter and then compute maxentropic density from the associated moments.

The data generation process

We shall consider $B = 3$ and $H_b = 7$ for all b. We shall consider several possible frequency models for the $N_{b,h}$ and, to create a first level of dependency, we shall couple them by means of Gaussian copulas with different values of the correlation coefficient. Below we describe the coupling procedure in more detail.

In Table 7.4, we list all the frequency models considered. To generate S_1 and S_2 we use almost the same inputs. S_2 consists of two pieces. One is a sample with the same distributions as, but independent of, S_1, and the other is an additional compound variable generated as explained in Table 7.5. The frequency was modeled by a negative binomial and the individual losses were chosen to be Pareto. This is done to add kurtosis to a sample like S_1. Not only that, we generated a very large amount of data, but in order to have relatively large numbers, we filtered out the losses smaller than 11 000 and retained over 9000 of those points. To fatten the tail behavior of S_2, from the Pareto data we filtered

Table 7.4: Inputs for the simulation of S_1.

S_{bh}	N_{bh}	X_{bh}
S_{11}:	POISSON ($\lambda = 20$)	CHAMPERNOWNE ($a = 25, M = 85, c = 15$)
S_{12}:	POISSON ($\lambda = 40$)	PARETO (shape = 10, scale = 85)
S_{13}:	BINOMIAL ($n = 70, p = 0.5$)	PARETO (shape = 10, scale = 85)
S_{14}:	BINOMIAL ($n = 62, p = 0.5$)	CHAMPERNOWNE ($a = 10, M = 100, c = 45$)
S_{15}:	BINOMIAL ($n = 50, p = 0.5$)	GAMMA (shape = 1500, rate = 15)
S_{16}:	BINOMIAL ($n = 56, p = 0.5$)	GAMMA (shape = 4000, rate = 35)
S_{17}:	NEGATIVE BINOMIAL ($r = 60, p = 0.3$)	WEIBULL (shape = 100, scale = 15)

Table 7.5: Additional inputs for the simulation of S_2.

	N_{bh}	X_{bh}
TAIL:	NEGATIVE BINOMIAL ($r = 90, p = 0.8$)	PARETO (shape = 5.5, scale = 2100)

out values smaller than 32 000 and again kept the same total number of points as for S_1 and S_2.

Let us now explain how we added dependency among the $S_{1,h}$. We supposed that there was some sort of dependence among the frequencies. This amounts to supposing that whatever it is that causes the dependence affects only the frequency at which events occur but not the individual severities. The frequencies shown in Table 7.4 proceed as follows: We use a copula with correlation $\rho = 0.8$ to couple $N_{1,1}$, $N_{1,2}$ and $N_{1,3}$, a copula of $\rho = 0.5$ to couple $N_{1,4}$, $N_{1,5}$ and a copula of $\rho = 0.3$ for $N_{1,6}$, $N_{1,7}$. And of course, a similar procedure applies to the part of S_2 generated similarly to S_1.

For S_3, we use the distributions described in Table 7.6, here we considered the compound variables $S_{3,h}$ for $h = 1, \ldots, 7$ as independent among themselves, and, as usual, the $N_{3,h}$ as independent of the individual losses $X_{3,h,k}$ with $k \geq 1$ independent and identically distributed. The size of the samples in each case are of 9000 data points.

Table 7.6: Inputs for the simulation of S_3.

S_{bh}	N_{bh}	X_{bh}
S_{31}:	POISSON ($\lambda = 80$)	CHAMPERNOWNE ($a = 20, M = 85, c = 15$)
S_{32}:	POISSON ($\lambda = 60$)	LogNormal ($\mu = -0.01, \sigma = 2$)
S_{33}:	BINOMIAL($n = 70, p = 0.5$)	PARETO (shape = 10, scale = 85)
S_{34}:	BINOMIAL ($n = 62, p = 0.5$)	CHAMPERNOWNE ($a = 10, M = 125, c = 45$)
S_{35}:	BINOMIAL ($n = 50, p = 0.5$)	GAMMA (shape = 4500, rate = 15)
S_{36}:	BINOMIAL ($n = 76, p = 0.5$)	GAMMA (shape = 9000, rate = 35)
S_{37}:	NEGATIVE BINOMIAL ($r = 80, p = 0.3$)	WEIBULL (shape = 200, scale = 50)
Tail:	NEGATIVE BINOMIAL ($r = 90, p = 0.8$)	PARETO (shape = 5.5, scale = 5550)

Once S_1, S_2, S_3 are obtained, we apply the maximum entropy technique as explained above to obtain their densities $f_1^*(s), f_2^*(s), f_3^*(s)$. These are plotted in Figure 7.4. The scale in the abscissa axis in all three plots is 10^5.

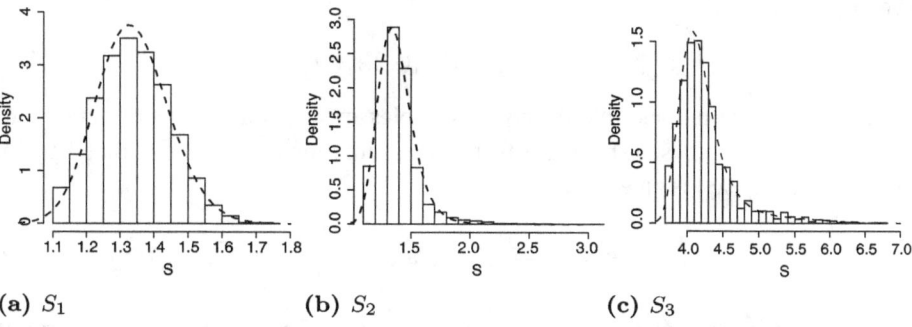

(a) S_1 (b) S_2 (c) S_3

Figure 7.4: Losses for each line of activity reconstructions reconstructed by SME.

The scale in each case is in thousands. This time we measure the quality of the reconstruction using the MAE and RSME distances between the reconstructed densities f_{S_i} and the histogram. We show the results in Table 7.7. The agreement is quite good.

Table 7.7: MAE and RMSE distances between the histogram and the S_i.

S	Error	
	MAE	RMSE
S_1	0.0054	0.0072
S_2	0.0241	0.0282
S_3	0.0061	0.0071

Once we have the $f_i^*(s)$ and the data, there are a variety of things that we can do to compute the distribution of the sum $S = S_1 + S_2 + S_3$.
(1) First, we can suppose that the S_i are independent, and use the data to generate a sample of $S = S_1 + S_2 + S_3$ to determine the distribution of total losses using SME.
(2) Use a sequential coupling combined with convolution to determine a density of total losses.
(3) There are several possible copulas that we can invoke to produce a joint distribution. Then we can sample from that distribution to generate joint samples (s_1, s_2, s_3) from which we can produce $s = s_1 + s_2 + s_3$ and then use SME to determine the density of total losses.

As the notion of copula is essential for what follows, we say a few words about the subject here and direct the interested reader to Chapter 14 for some more about the subject. A copula is the joint distribution function of $n \geq 1$ random variables $Y_1, \ldots, Y_n$ that take values in $[0, 1]$. We shall denote it by the customary $C(u_1, \ldots, u_n): [0, 1]^n \to [0, 1]$, and suppose that it is continuously differentiable in all variables. Given a copula (there is a large catalog of them), if one knows the cumulative distribution functions $F_i(x_i)$ of the random variables $X_1, \ldots, X_n$, then a candidate for joint distribution for $(X_1, \ldots, X_n)$ is given by

$$F_{X_x,\ldots,X_n}(x_1,\ldots,x_n) = C(F_1(x_1),\ldots,F_n(x_n)). \tag{7.22}$$

If, furthermore, we suppose that the given variables are continuous with densities $f_i(x_i)$ with respect to the Lebesgue measure, from our assumption on C it follows that the joint density of $(X_1, \ldots, X_n)$ is given by

$$f_{X_1,\ldots,X_n}(x_1,\ldots x_n) = \frac{\partial^n}{\partial u_1,\ldots,\partial u_n} C(u_1,\ldots u_n) \Big|_{u_i=F_i(x_i)} \prod_{i=1}^n f_i(x_i). \tag{7.23}$$

Now we come back to our problem. For us, $n = 3$ and the $f_i(x_i)$ are the maxentropic densities of the S_i that we determined above. Once we settle on a copula, we generate a sample (s_1, s_2, s_3) from the joint distribution given by (7.22), and for each sample point we form $s = s_1 + s_2 + s_3$, which will be the inputs for the determination of the maxentropic density of the total aggregate loss, then carry on the standard business, that is, we implement the maxentropic procedure and perform a few comparisons.

To take care of the first issue in the itemized list a few lines above we consider the independent copula $C(u_1, u_2, u_3) = u_1 u_2 u_3$, which amounts to saying that we consider the (S_1, S_2, S_3) as independent and just generate $s = s_1 + s_2 + s_3$ from the data that we already have. In this case, since there are only three densities, the computations to determine the joint density by convolution are not that heavy, namely

$$f_S(s) = \int_0^s f_1(s - s_1) \left(\int_0^{s_1} f_2(s_1 - s_2) f_3(s_2) ds_2 \right) ds_1. \quad (7.24)$$

For the second item, we shall suppose that there is dependence between S_1 and S_2 modeled by a copula $c(u_1, u_2)$ and then suppose that $L_{1,2} = S_1 + S_2$ is independent of S_3. To obtain the total loss in two steps, we first obtain f_{S_1, S_2} according to (7.23), and then generate a sample for $L_{1,2}$ with density

$$f_{L_{1,2}}(\ell) = \int_0^\ell f_{S_1, S_2}(s, \ell - s) ds. \quad (7.25)$$

This sequential procedure seems to have been first introduced by Alexander in [5]. With this and the sample for S_3 we generate a sample for $s = \ell + s_3$, which is then used as input for the maximum entropy procedure. Note that also in this case the density of the total loss can be obtained by convolution of $f_{L_{1,2}}$ and f_3, namely

$$f_S(s) = \int_0^s f_{L_{1,2}}(s - \ell) f_3(\ell) d\ell. \quad (7.26)$$

This and the three term convolution will be used below to compare the output of the maxentropic procedure when the data is generated according to the assumed densities.

Finally, for the third item in the list we consider a pair of copulas from the catalog and generate the necessary data from the resulting joint density. Let us now describe the numerical results and some comparisons among them.

Independent copula versus partial coupling

Let us first compare the results provided by the maxentropic techniques when the partial losses are either independent or only the first two of them are dependent but independent of the third. The copulas that were used to construct a joint density of (S_1, S_2) were the following (see Figure 7.5). To generate the data for panel (a) we supposed that

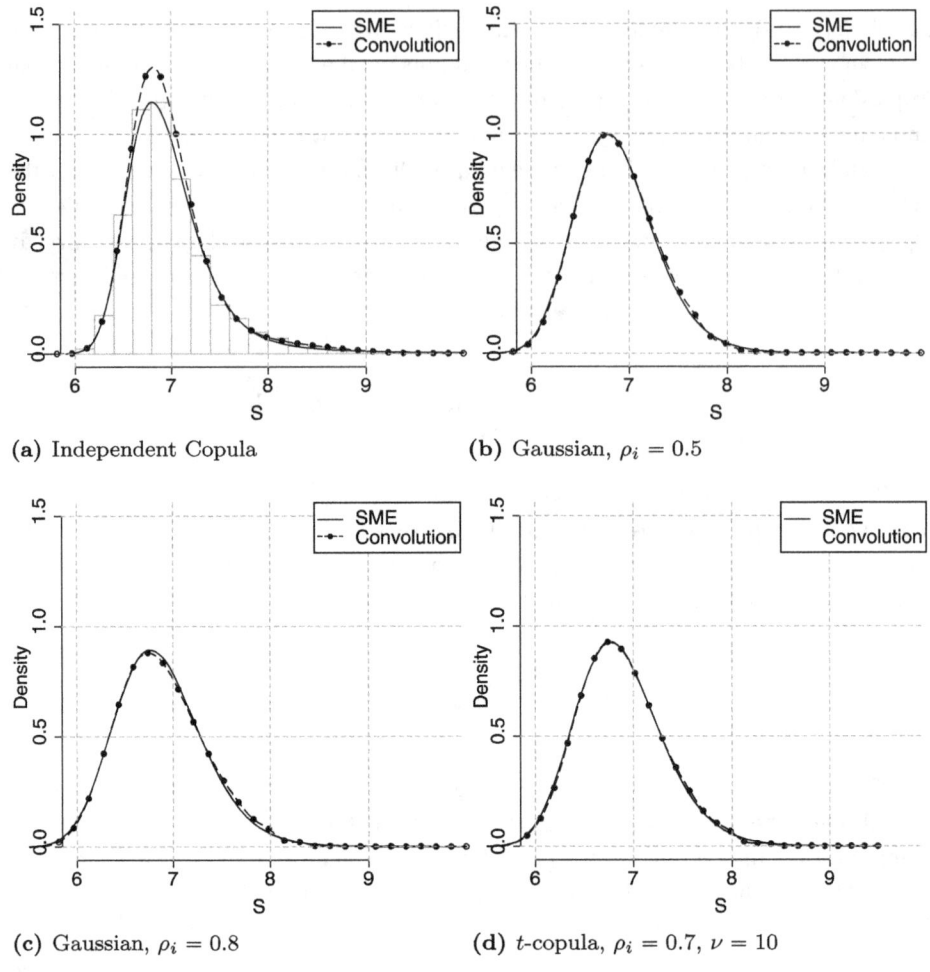

Figure 7.5: Density of the sum obtained by convolution and SME approaches for different types of copulas and correlations.

S_1, S_2 and S_3 were independent. For the data for (b)–(c) we used Gaussian copulas with $\rho = 0.5$ and $\rho = 0.8$, whereas to generate the data in panel (d) we used a Student copula with parameters $\rho_i = 0.7$, $\nu = 10$. This one is usually used to generate long tailed data.

The results are plotted in Figure 7.5 along with the results of calculating the densities by direct convolution.

The contents of the four panels in Figure 7.5 are as follows: In panel (a) we show the result of applying the maxentropic procedure to the empirical data obtained by simply summing the samples of each of the losses as well as the result of obtaining the density numerically by convolution according to (7.24). In panels (b)–(d) we show the maxentropic density obtained when the input data was $s = \ell_{1,2} + s_3$ where $\ell_{1,2}$ was sampled from the joint F_{S_1,S_2} couple by the copula indicated, and regarding S_3 independent of that partial total loss. There we also show the density of total loss computed as indicated in (7.26).

To compare the maxentropic density to the true density we measured the discrepancy between them using the MAE and the RSME. The results are displayed in Table 7.8. These results confirm the visual results apparent in Figure 7.5.

Table 7.8: MAE and RMSE distances between SME and the convolution approaches.

Copula	Error	
	MAE	RMSE
Independent	0.027	0.039
Gaussian, $\rho = 0.5$	0.004	0.005
Gaussian, $\rho = 0.8$	0.004	0.005
t-Student, $\rho = 0.7, v = 10$	0.004	0.005

Fully dependent risks

The numerical experiments in this section go as follows. We shall consider a few copulas that will provide us with a joint density for (S_1, S_2, S_3). Recall that above we mentioned that since the maximum entropy method provides us with reliable partial loss distributions, we can use them to determine a joint distributions or joint density as in (7.22) and (7.23). So, the problem of determining a density of the total loss will be solved, because from that one can compute the distribution of $S_1 + S_2 + S_3$. The actual computation goes as follows. From the joint density f_{S_1,S_2,S_3} one can compute the probability (cumulative distribution function) by

$$P(S_1 + S_2 + S_3 \leq x) = \int_{\{s_1+s_2+s_3 \leq x\}} f_{S_1,S_2,S_3}(s_1, s_2, s_3) ds_1 ds_2 ds_3,$$

which after the change of variables $s_1 + s_2 + s_3 = s$, $s_2 + s_3 = u$, $s_3 = v$, differentiation with respect to x and renaming variables, becomes

$$f_S(s) = \int_0^s \left(\int_0^u f_{S_1,S_2,S_3}(s-u, u-v, v) dv \right) du. \quad (7.27)$$

This is the 'exact' density against which we shall compare the output of the maxentropic procedure. Of course, for the comparison to be fair, the input data has to be well collected, that is, we suppose that $s = s_1 + s_2 + s_3$ is obtained from a well collected sample, that is a sample generated from the true distribution F_{S_1,S_2,S_3} obtained by coupling according to (7.22).

For the numerical examples we considered a several copulas. For the first example we considered a Gumbel and a Clayton copula, with parameters 3 and 2 respectively. We carried out the maxentropic approach with the data provided by the Gumbel copula, and computed the 'exact' densities according to (7.27) for the densities obtained from the Gumbel and Clayton copulas. The result obtained is plotted in Figure 7.6. In the plot we also show the histogram of the data.

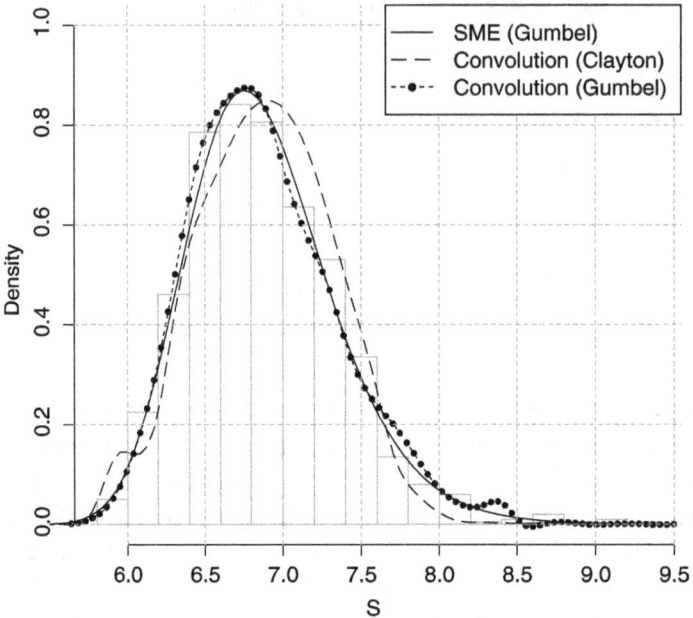

Figure 7.6: Densities for the Gumbel and Clayton copulas.

The plot may be read as follows. First, to exemplify the fact that when the data is well collected (that is, the different random variables are collected jointly, or in such a way that the dependencies are preserved), the maximum entropy method provides a reliable reconstruction of the true density. Second, we exemplify the effect of the choice of the wrong copula to determine the density of the total loss, and finally, the effect of the numerical convolution procedure, which may be the cause of the wiggle in the curves obtained as in (7.27).

As a measure of the quality of the reconstruction in Table 7.9, we consider the MAE and RMSE between the histogram and the three other curves. In this experiment, the maxentropic reconstruction has the smallest error. In [46] we present further examples of a similar analysis carried out with Gaussian copulas with positive and negative correlation and one obtained with t-Student based copula. Also, in Chapter 12, where we carry out computations made with the maxentropic densities, we present the bearing

Table 7.9: Errors of SME versus convolution approaches.

Error	SME	Convolution (Gumbel)	Convolution (Clayton)
MAE	0.005107	0.005453	0.02512
RMSE	0.006291	0.007283	0.03088

of the copula choice on the computation of quantiles and tail conditional expectations, that is, the influence of the choice of the copula on the computation of regulatory capital.

7.5.3 A comparison of methods

This section is devoted to a comparison of the four different methods in a case in which an analytic solution to the problem is at hand. The model we shall consider for the compound losses is the following: The frequency is to be modeled by a Poisson distributed random variable and the individual losses are supposed to be distributed according to a gamma density. In this case, we saw in Chapter 5 that the exact densities for any number of losses are again gamma distributions and that the problem boils down to a summation of an infinite series of gamma densities.

Not only that, but the Laplace transform of the density of the compound variable can be explicitly computed, and different methods of inverting it can be compared. We consider the maxentropic method, which depends on a small number of data points versus an integer moment based method and a Fourier inversion method, which in this case is easy to implement because the Laplace transform is simple to extend to the right half complex plane.

If the intensity parameter of the Poisson is denoted by ℓ and the individual losses have densities $\Gamma(a, b)$, we saw that the density of the total loss (given that losses occur) is given by

$$f_S(x) = \frac{e^{-\ell} e^{-bx}}{1 - e^{-\ell}} \sum_{n=1}^{\infty} \frac{\ell^n x^{na-1} b^{na}}{n! \Gamma(na)}.$$

To chose a truncation point for the sum, we determine an n_ℓ such that

$$\frac{e^{-\ell}}{1 - e^{-\ell}} \sum_{n=1}^{\infty} \frac{\ell^n}{n!} \geq 1 - \epsilon$$

for $\epsilon = 10^{-5}$. For the numerical example developed in [50], from which we took Figure 7.7, the parameters chosen were $\ell = 1$, $a = 2$ and $b = 1$ and two ways of assigning the eight Laplace transform parameters α were considered.

Note that the maxentropic method and the Fourier inversion method give quite good approximations to the (approximate) true density obtained by summing $n_\ell = 500$

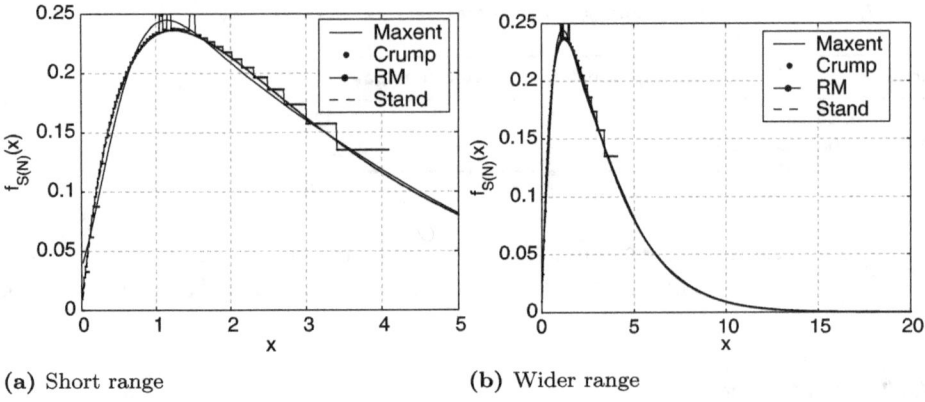

(a) Short range (b) Wider range

Figure 7.7: A comparison of density reconstruction methods.

terms. The difference being of course that to implement the Fourier summation we need to have an analytic expression for the Laplace transform. Similarly, for the integer based method we need a large number of integer moments whereas for the maxentropic method we need only eight fractional moments. Again, consider Chapter 9 for an explanation of this phenomenon.

To conclude we should mention an important numerical issue that comes up when minimizing the dual entropy function to determine the Lagrange multipliers. The function $\Sigma\lambda$ is in most cases quite flat near the minimum. Thus to find the minimum when using gradient type methods it is convenient to couple such methods with a step reducing procedure. Such a method exists, and it is called the Barzilai and Borwein algorithm (BB-method), which is a nonmonotone gradient method. This is a low-cost computational method, easy to implement and very effective for large-scale problems [83].

8 Extensions of the method of maximum entropy

We begin this chapter by explaining the method of maximum entropy in the mean through its application to solve a linear ill-posed inverse problem with box constraints. The same framework will be implicit in the extensions of the method that we consider afterward.

The statement of the problem that we consider in the first section goes as follows:

$$\text{Determine} \quad x \in \mathcal{C} \quad \text{such that} \quad Ax = y \quad \text{for } y \in \mathbb{R}^d. \tag{8.1}$$

Here $\mathcal{C} = \prod_{i=1}^{n}[L_i, U_i]$ with $-\infty < U_i < L_i < +\infty$ for $i = 1, \ldots, n$, and $d < n$. The procedure developed in this section will be used to deal with the capital allocation problem.

The two extensions of the maximum entropy method are designed to cover two somewhat related cases:
(1) There may be errors in the data. This means that the moments μ_i that appear in the right-hand side of (7.1)–(7.2) are exactly known, but up to errors in the measurement.
(2) The moments μ_i are not known. What is known is an interval, say $[a_i, b_i]$, in which the moment μ_i falls.
(3) There is error in the measurement, and the data is specified to lie in an interval.

Let us state the three possible generalized moment problems. The first one consists of finding the probability density f of an unknown probability P with respect to a reference measure m on a sample space $(\Omega, \mathcal{F})$ and on estimating a K-vector $\boldsymbol{\eta}$ such that

$$\int_\Omega h_i(X(\omega))f(\omega)dm(\omega) + \eta_i = \mu_i, \quad i = 1, \ldots, K. \tag{8.2}$$

Observe that in this problem the error η_i is known to be there but is an unknown quantity that has to be estimated as well. At this point, some modeling is necessary to further specify the problem. This will be done below. We will maintain a general notation for a while but eventually replace Ω by $[0, \infty)$, P by F_X, etc., when we come down to our specific application.

The second problem can be stated as consisting of finding a density f such that

$$\int_\Omega h_i(X(\omega))f(\omega)dm(\omega) \in [a_i, b_i], \quad i = 1, \ldots, K, \tag{8.3}$$

where the $a_i, b_i, i = 1, \ldots, K,$ are to be provided by the modeler. Here we suppose that the left-hand side can be computed exactly, but its value is specified or known to lie in a given interval. This way of looking at the problem is natural in some cases in which a bid-ask range for the price of a risk or an asset is known, and we are interested in determining the probability distribution implied in that price. The case in which the expected values of the X_i are estimated by "experts" to lie in ranges is also typical.

To finish, we also have the problem in which not only there is a possible measurement error in the evaluation or determination of the left-hand side of (7.2), but also the data are specified up to a range. That is, we want to find a density f and a vector $\boldsymbol{\eta}$ such that

$$\int_\Omega h_i(\boldsymbol{X}(\omega))f(\omega)dm(\omega) + \eta_i \in [a_i, b_i], \quad i = 1, \ldots, K. \tag{8.4}$$

The work in this chapter is based on [47], although there are proposals different from ours to infer probabilities from data in ranges, like that of [1].

8.1 Generalized moment problem with errors in the data

Let us consider (8.2). If we are to think of the μ_i as expected values of $h_i(\boldsymbol{X})$, that is, as possible average values over some sample values, then it is consistent to think of the error in the specification of μ_i as the average of the errors incurred in the observations of the X_i themselves or perhaps errors in the numerical estimation of the expected values. That is, we may suppose that

$$\eta_i = \int_{\Omega_n} y_j f_n(\boldsymbol{y}) dv(\boldsymbol{y}),$$

where the model for the sample space $(\Omega_n, \mathcal{F}_n)$ has to be specified as well as that of the measure dv. There are two extreme cases to consider for the range of the random variables Y_i modeling the measurement error. Either they have a finite range, that is, $Y_i \in [-\delta_i, \delta_i]$, where the δ_i are to be provided by the modeler, or they have an infinite range, that is, $Y_i \in \mathbb{R}$. Let us consider simple versions of both cases separately.

8.1.1 The bounded error case

Notice now that as any number in $[-\delta_i, \delta_i]$ can be written as a convex combination of the two end points, it suffices to consider measures that are absolutely continuous with respect to

$$dv(\boldsymbol{y}) = \prod_{i=1}^{K}(\epsilon_{-\delta_i}(dy_i) + \epsilon_{\delta_i}(dy_i)),$$

where $\epsilon_a(dy)$ denotes the unit point mass concentrated at the point a, also called the Dirac measure concentrated at a. Observe that any probability P_n on $\Omega_n = \prod_{i=1}^{K}[-\delta_i, \delta_i]$ absolutely continuous with respect to $dv(\boldsymbol{y})$ has the form

$$dP_n(y) = \prod_{i=1}^{K}(p_i\epsilon_{-\delta_i}(dy_i) + (1-p_i)\epsilon_{\delta_i}(dy_i)),$$

and the expected value of any Y_i is given by

$$E_{P_n}[Y_i] = -p_i\delta_i + (1-p_i)\delta_i.$$

Having said this, our problem consists of finding a product probability $dP(x)dP_n(y)$ on $\Omega \times \Omega_n$ such that

$$E_{P\times P_n}[h_i(X) + Y_i] = \int h_i(X)dP + \int Y_i dP_n = \mu_i. \tag{8.5}$$

As it is easy to see that the entropy of a product measure is the sum of the entropies of the factor measures, our problem now is finding a density $f(x)$ and numbers $0 < p_i < 1$ such that

$$S_m(f,\boldsymbol{p}) = S_m(f) - \sum_{i=1}^{K}(p_i \ln p_i - (1-p_i)\ln(1-p_i))$$

achieves its maximum values over the class satisfying (8.5). The procedure for this case is the same as above, except that now the function $Z(\lambda)$ has more structure built into it. This time,

$$Z(\lambda) = \int e^{-\langle \lambda, h(X) \rangle} dm \times \prod_{i=1}^{K}(e^{\lambda_i \delta_i} + e^{-\lambda_i \delta_i}), \tag{8.6}$$

and the dual entropy function remains the same, that is,

$$\Sigma(\lambda, \boldsymbol{\mu}) = \ln Z(\lambda) + \langle \lambda, \boldsymbol{\mu} \rangle = \ln Z_X(\lambda) + \sum_{i=1}^{K}\ln(e^{\lambda_i \delta_i} + e^{-\lambda_i \delta_i}) + \langle \lambda, \boldsymbol{\mu} \rangle,$$

except that $Z(\lambda)$ is given by the expression displayed two lines above, and we now use $Z_X = \int e^{-\langle \lambda, h(X) \rangle} dm$. The result that we need looks similar to Theorem 7.2, and we state it next.

Theorem 8.1. *With the notations introduced above, under Assumption A, suppose that the minimum of $\Sigma(\lambda, \boldsymbol{\mu})$ as function of λ is achieved at λ^* in the interior of $\mathcal{D}(m)$. Then*

$$-\nabla_\lambda \ln Z(\lambda^*) = \boldsymbol{\mu}, \tag{8.7}$$

the density

$$\rho_{\lambda^*} = \frac{e^{-\langle \lambda^*, h(X) \rangle}}{Z(\lambda^*)} \tag{8.8}$$

and the probability

$$p_i^* = \frac{e^{\delta_i \lambda_i^*}}{e^{\delta_i \lambda_i^*} + e^{-\delta_i \lambda_i^*}} \tag{8.9}$$

solve the entropy maximization problem, and furthermore

$$S(\rho_{\lambda^*}) = \Sigma(\lambda^*, \mu) = \ln Z(\lambda^*) + \langle \lambda^*, \mu \rangle. \tag{8.10}$$

Comment. The proof proceeds like that of Theorem 7.2. We only add that the estimate of the observational error in this model is given by

$$e_i^* = -\delta_i \frac{e^{\delta_i \lambda_i^*}}{e^{\delta_i \lambda_i^*} + e^{-\delta_i \lambda_i^*}} + \delta_i \frac{e^{-\delta_i \lambda_i^*}}{e^{\delta_i \lambda_i^*} + e^{-\delta_i \lambda_i^*}}.$$

8.1.2 The unbounded error case

The underlying logic in this case is the same as in the previous case, except that now we suppose that as a reference measure on the error sample space we consider a Gaussian measure, that is,

$$d\nu(\mathbf{y}) = \prod_{i=1}^{K} e^{-y^2/2} \frac{d\mathbf{y}}{(2\pi)^{K/2}}.$$

A probability absolutely continuous with respect to this reference measure has a density $f_n(\mathbf{y})$, which will be required to satisfy (8.5) with $dP_n(\mathbf{y}) = f_n(\mathbf{y}) d\nu(\mathbf{y})$. The entropy function of the product density $f(\omega) f_n(\mathbf{y})$ is

$$S_{m \times \nu}(f, f_n) = S_m(f) + S_\nu(f_n).$$

Again, instead of maximizing the entropy, we minimize the dual entropy, and for that, the first step is to compute the normalization factor $Z(\lambda)$. A simple calculation shows that

$$Z(\lambda) = \int e^{-\langle \lambda, \mathbf{h}(X) \rangle} dm \times \prod_{i=1}^{K} e^{\|\lambda\|^2/2}. \tag{8.11}$$

Therefore the dual entropy is given by

$$\Sigma(\lambda, \mu) = \ln Z(\lambda) + \langle \lambda, \mu \rangle = \ln Z_X(\lambda) + \frac{1}{2} \sum_{i=1}^{K} \lambda_i^2 + \langle \lambda, \mu \rangle.$$

There is a slight change in notation in the statement of the main result for this case.

Theorem 8.2. *With the notations introduced above, under Assumption A, suppose that the minimum of $\Sigma(\lambda, \mu)$ as function of λ is achieved at λ^* in the interior of $\mathcal{D}(m)$. Then*

$$-\nabla_\lambda \ln Z(\lambda^*) = \mu, \tag{8.12}$$

the densities

$$\rho_{\lambda^*} = \frac{e^{-\langle \lambda^*, h(X) \rangle}}{Z(\lambda^*)} \tag{8.13}$$

and

$$f_n(y)^* \prod_{i=1}^{K} \frac{1}{\sqrt{2\pi}} e^{-\|y\|^2/2} = \prod_{i=1}^{K} \frac{1}{\sqrt{2\pi}} e^{-\|y+\lambda^*\|^2} \tag{8.14}$$

solve the entropy maximization problem, and furthermore

$$S_{m \times \nu}(\rho_{\lambda^*} f_n^*) = \Sigma(\lambda^*, \mu) = \ln Z_X(\lambda^*) + \frac{1}{2}\|\lambda^*\|^2 + \langle \lambda^*, \mu \rangle. \tag{8.15}$$

8.1.3 The fractional moment problem with bounded measurement error

As an application of the first case, we consider the fractional moment problem considered in Section 7.4.1 when a measurement error is to be taken into account. We already explained that to do that, we have to modify the notations appropriately. We pointed out that in this case, $\Omega = [0, 1]$ and $\mathcal{F} = \mathcal{B}([0,1])$, that is, $dm = dy$. Instead of X, we have $Y(y) = y$ and $h_i(y) = y^{\alpha_i}$. According to (8.1), the maxentropic density is given by

$$\rho_{\lambda^*} = \frac{e^{-\langle \lambda^*, h(y) \rangle}}{Z(\lambda^*)}, \tag{8.16}$$

where $Z(\lambda)$ is detailed in Section 7.4.1, and the errors are to be estimated as

$$\epsilon_i^* = -\delta_i \frac{e^{\delta_i \lambda_i^*}}{e^{\delta_i \lambda_i^*} + e^{-\delta_i \lambda_i^*}} + \delta_i \frac{e^{-\delta_i \lambda_i^*}}{e^{\delta_i \lambda_i^*} + e^{-\delta_i \lambda_i^*}}.$$

Also, (8.10) holds, that is,

$$S(\rho_{\lambda^*} f_n^*) = \Sigma(\lambda^*, \mu) = \ln Z(\lambda^*) + \langle \lambda^*, \mu \rangle,$$

where now $Z(\lambda)$ is given by

$$Z(\lambda) = \int_0^1 e^{-\sum_{i=1}^K \lambda_i y^{\alpha_i}} dy \prod_{i=1}^K (e^{\lambda_1 \delta_i} + e^{-\lambda_1 \delta_i}).$$

Note that the normalization factor is finite for all $\lambda \in \mathbb{R}^K$.

8.2 Generalized moment problem with data in ranges

This section is devoted to the solution of problem (8.3), namely finding a density (with respect to dm) such that

$$\int_\Omega h_i(X(\omega))f(\omega)dm(\omega) \in [a_i, b_i], \quad i = 1, \ldots, K,$$

where the $a_i, b_i, i = 1, \ldots, K$, must be specified as part of the statement of the problem. The intuition behind the method that we propose next goes as follows. For each $\mu_i \in [a_i, b_i]$, we solve the maximum entropy problem. The entropy of the resulting density depends on μ; denote it by $S_m(\mu)$. Then we determine the μ that maximizes the set $\{S_m(\mu) \mid \mu \in \prod_{i=1}^K [a_i, b_i]\}$.

Let us put

$$\mathcal{P}(\mu) = \left\{ \text{Densities } f \text{ such that (7.2) holds for } \mu \in \prod_{i=1}^K [a_i, b_i] \right\}$$

and

$$\mathcal{P} = \{\text{Densities } f \text{ such that (8.3) holds}\} = \bigcup_{\mu \in \prod_{i=1}^K [a_i, b_i]} \mathcal{P}(\mu).$$

Then we clearly have

$$\sup\{S_m(f) \mid f \in \mathcal{P}\} = \sup\left\{\sup\{S_m(f) \mid f \in \mathcal{P}(\mu)\} \,\middle|\, \mu \in \prod_{i=1}^K [a_i, b_i] \right\}.$$

Now recall from (8.10) in Theorem 7.2 that at the density at which $\sup\{S_m(f) \mid f \in \mathcal{P}(\mu)\}$ is reached, we have

$$S_m(\mu) = \inf\{\ln Z(\lambda) + \langle \lambda, \mu \rangle \mid \lambda \in \mathcal{D}(m)\},$$

which we substitute in the previous identity to obtain

$$\sup\{S_m(f) \mid f \in \mathcal{P}\} = \sup\left\{\inf\{\ln Z(\lambda) + \langle \lambda, \mu \rangle \mid \lambda \in \mathcal{D}(m)\} \,\middle|\, \mu \in \prod_{i=1}^K [a_i, b_i] \right\}.$$

Recall that under Assumption A, we have the following:

Lemma 8.1. *The function* $\Sigma(\lambda, \mu): \mathcal{D}(m) \times \mathbb{R}^K \to \mathbb{R}$ *given by*

$$(\lambda, \mu) \to \ln Z(\lambda) + \langle \lambda, \mu \rangle$$

is convex in λ and concave in μ, and in the interior of $\mathcal{D}(m) \times \mathbb{R}^K$, we have

$$\nabla_\mu \nabla_\lambda \Sigma(\lambda, \mu) = \mathbf{I}.$$

Here **I** is the $K \times K$ identity matrix, which is clearly a positive definite matrix. The proof of this fact is trivial. A consequence of the lemma is that the maximum and minimum in the last identity of the chain can be exchanged, and we have

$$\sup\{S_m(f) \mid f \in \mathcal{P}\} = \inf\left\{\sup\left\{\ln Z(\lambda) + \langle \lambda, \mu \rangle \,\Big|\, \mu \in \prod_{i=1}^{K}[a_i, b_i]\right\} \,\Big|\, \lambda \in \mathcal{D}(m)\right\},$$

which can be restated as

$$\sup\{S_m(f) \mid f \in \mathcal{P}\} = \inf\left\{\ln Z(\lambda) + \sup\left\{\langle \lambda, \mu \rangle \,\Big|\, \mu \in \prod_{i=1}^{K}[a_i, b_i]\right\} \,\Big|\, \lambda \in \mathcal{D}(m)\right\}.$$

To compute the inner maximum, we begin with the following:

Lemma 8.2. *For $\lambda \in \mathbb{R}^K$, we have*

$$\sup\{\langle \lambda, \mu \rangle \mid \mu \in [-1, 1]^K\} = \|\lambda\|_1 = \sum_{i=1}^{K} |\lambda_i|.$$

To complete the maximization, we map $[-1, 1]^K$ homotetically onto $\prod_{i=1}^{K}[a_i, b_i]$ by means of $\xi \to \mu = \kappa + \mathbf{T}(\xi)$, where $\kappa_i = (a_i + b_i)/2$, and $\mathbf{T}$ is diagonal with entries $T_{ii} = (b_i - a_i)/2$. Therefore

$$\langle \lambda, \mu \rangle = \langle \lambda, \kappa \rangle + \langle \mathbf{T}\lambda, \xi \rangle,$$

and it follows from Lemma 8.2 that

$$\sup\left\{\langle \lambda, \mu \rangle \,\Big|\, \mu \in \prod_{i=1}^{K}[a_i, b_i]\right\} = \langle \lambda, \kappa \rangle + \|\mathbf{T}\lambda\| = \sum_{i=1}^{K}\left(\lambda_i \frac{(a_i + b_i)}{2} + |\lambda_i|\frac{(b_i - a_i)}{2}\right). \quad (8.17)$$

Therefore the chain of identities becomes

$$\sup\{S_m(f) \mid f \in \mathcal{P}\} = \inf\{\ln Z(\lambda) + \langle \lambda, \kappa \rangle + \|\mathbf{T}\lambda\| \mid \lambda \in \mathcal{D}(m)\}.$$

Due to the presence of the absolute values in (8.17), to state an analogue of Theorem 7.2, we have to recall the extension of the notion of derivatives. As motivation, note that the slope of the standard absolute value function $|x|$ at 0 is any number in $[-1, 1]$, which are the possible slopes of the tangent lines to the function $|x|$ at 0. The generic definition goes as follows (see [12]).

Definition 8.1. Let $f: \mathcal{D} \to (0, \infty]$ be a convex function defined on a convex domain in $\mathbb{R}^K$. The subgradient $\partial f(x_0)$ is the *set* of all vectors $\xi \in \mathbb{R}^K$ such that

$$\langle \xi, x - x_0 \rangle \leq f(x) - f(x_0).$$

If $x_0 \notin \mathcal{D}$, then we set $\partial f(x_0) = \emptyset$.

The subgradient is a closed convex set. The important result that we need is the following:

Proposition 8.1. *Suppose that the domain $\mathcal{D}$ has a nonempty interior. Then $x_0 \in \mathcal{D}$ is a global minimizer of the convex function $f: \mathcal{D} \to (0, \infty]$ if and only if the extended first-order condition*

$$0 \in \partial f(x_0)$$

holds.

We can now state the analogue of Theorem 7.2.

Theorem 8.3. *Suppose that the function $\ln Z(\lambda) + \langle \lambda, \kappa \rangle + \|\mathbf{T}\lambda\|$ achieves its minimum at λ^*. At this minimum, we have*

$$\begin{aligned}
-\frac{\partial \ln Z(\lambda^*)}{\partial \lambda_i} &= b_i & \text{if } \lambda_i^* > 0, \\
-\frac{\partial \ln Z(\lambda^*)}{\partial \lambda_i} &\in [a_i, b_i] & \text{if } \lambda_i^* = 0, \\
-\frac{\partial \ln Z(\lambda^*)}{\partial \lambda_i} &= a_i & \text{if } \lambda_i^* < 0.
\end{aligned} \quad (8.18)$$

In this case the maxentropic density has the usual representation

$$\rho_{\lambda^*} = \frac{e^{-\langle \lambda^*, X \rangle}}{Z(\lambda^*)}, \quad (8.19)$$

which solves the entropy maximization problem, and furthermore

$$S(\rho_{\lambda^*}) = \ln Z(\lambda^*) + \langle \lambda^*, \kappa \rangle + \|\mathbf{T}\lambda^*\|. \quad (8.20)$$

8.2.1 Fractional moment problem with data in ranges

This time, we consider another variation on the fractional moment problem addressed in Section 7.4.1, except that now we suppose that the data is known up to a range. We will particularize for this situation the results just obtained. Now, as in Section 8.1.3, we have only one random variable that takes values in $[0, 1]$ and K functions $h_i(y) = y^{\alpha_i}$ for $i = 1, \ldots, K$. Therefore the setup will be as in that section. The problem that we consider now consists of finding a density $f(y)$ on $[0, 1]$ such that

$$\int_0^1 y^{\alpha_i} f(y) dy \in [a_i, b_i], \quad i = 1, \ldots, K.$$

We saw that the solution to this problem can be represented by

$$f^*(y) = \frac{e^{-\langle \lambda^*, h(y) \rangle}}{Z(\lambda)},$$

where the vector λ^* is obtained by minimizing the dual entropy

$$\Sigma(\lambda) = \ln Z(\lambda) + \sum_{i=1}^{K}\left(\lambda_i \frac{(a_i + b_i)}{2} + |\lambda_i|\frac{(b_i - a_i)}{2}\right)$$

with

$$Z(\lambda) = \int_0^1 e^{-\langle \lambda^*, h(y) \rangle} dy.$$

The normalization factor is again finite for $\lambda \in \mathbb{R}^K$, and the minimizer differs from that of the previous example because the dual entropy is different.

8.3 Generalized moment problem with errors in the data and data in ranges

This section is devoted to problem (8.4), that is, to determining a density f such that

$$\int_\Omega h_i(X(\omega))f(\omega)dm(\omega) + \eta_i \in [a_i, b_i], \quad i = 1, \ldots, K. \tag{8.21}$$

In this problem, the data ranges $[a_i, b_i]$ and the nature of the measurement error are provided by the model builder. Its should be clear that to solve this problem by means of the method of maximum entropy, we just have to put together the two techniques developed in Sections 8.1 and 8.2.

To take care of the measurement error, we had to augment the sample space and extend the entropy to the class of densities on the extended space, and the rest of the procedure is as described in Chapter 7.

To take care of the data in ranges, we proceeded through a nested double maximization process, which eventually resulted in the minimization of a more elaborate dual entropy function.

The notations and the assumptions made above will stay in force now. We will consider only the case in which the measurement error is bounded. Let us first recall the normalization factor given by (8.6)

$$Z(\lambda) = \int e^{-\langle \lambda, h(X) \rangle} dm \times \prod_{i=1}^{K}(e^{\lambda_i \delta_i} + e^{-\lambda_i \delta_i}). \tag{8.22}$$

This normalization factor is to be incorporated into the dual entropy function given by (8.20) to obtain

$$\Sigma(\lambda) = \int e^{-\langle \lambda, h(X) \rangle} dm \times \prod_{i=1}^{K} (e^{\lambda_i \delta_i} + e^{-\lambda_i \delta_i}) + \sum_{i=1}^{K} \left(\lambda_i \frac{(a_i + b_i)}{2} + |\lambda_i| \frac{(b_i - a_i)}{2} \right). \quad (8.23)$$

The formal result can be stated as follows.

Theorem 8.4. *Suppose that the dual entropy has a minimizer λ^* lying in the interior of its domain. Then, the maximum entropy density f^* and weights p_i^* that solve (8.21) are given by*

$$f^* = \frac{e^{-\langle \lambda^*, h(X) \rangle}}{Z(\lambda^*)} \quad (8.24)$$

and

$$p_i^* = \frac{e^{\delta_i \lambda_i^*}}{e^{\delta_i \lambda_i^*} + e^{-\delta_i \lambda_i^*}}. \quad (8.25)$$

8.4 Numerical examples

Here we consider two variants of the example considered in Section 7.5, namely the problem of reconstructing a density from its Laplace transform regarded as a fractional moment problem, except that now we suppose that the data are only given in an interval or they are noisy data. We generate the data as in Section 7.5.

Recall that we considered two samples from a lognormal density with parameters $\mu = 1$ and $\sigma = 0.1$ of sizes $N = 200$ and $N = 1000$. The moments of these samples are given in Table 8.1. The corresponding confidence intervals are listed in Table 8.2.

Table 8.1: Moments of S for different sample sizes.

Size	Moments of S							
	μ_1	μ_2	μ_3	μ_4	μ_5	μ_6	μ_7	μ_8
200	0.0176	0.1304	0.2562	0.3596	0.4409	0.5051	0.5568	0.5990
1000	0.0181	0.1319	0.2579	0.3612	0.4424	0.5066	0.5581	0.6002

These will be the input for the first of the maxentropic reconstruction that we will carry out below and were computed in the standard way, that is, using the confidence interval

$$\left(\bar{\mu}_i - z^* \frac{sd_i}{\sqrt{n}}, \bar{\mu}_i + z^* \frac{sd_i}{\sqrt{n}} \right) \quad (8.26)$$

Table 8.2: Confidence intervals of $\bar{\mu}$ for different sample sizes.

μ_i	Sample size 200	Sample size 1000
μ_1	[0.0175, 0.0177]	[0.0181, 0.0182]
μ_2	[0.1302, 0.1307]	[0.1318, 0.1321]
μ_3	[0.2559, 0.2565]	[0.2578, 0.2581]
μ_4	[0.3592, 0.3599]	[0.3611, 0.3614]
μ_5	[0.4405, 0.4412]	[0.4423, 0.4426]
μ_6	[0.5048, 0.5054]	[0.5065, 0.5068]
μ_7	[0.5565, 0.5570]	[0.5580, 0.5583]
μ_8	[0.5987, 0.5992]	[0.6001, 0.6004]

with upper $(1 - C)/2$ critical value and lower $1 - (1 - C)/2$ critical value at confidence level C. For our example, we considered a confidence level of $C = 10\,\%$. In this case the confidence interval is $(\bar{\mu}_i - z_{0.55} sd_i/\sqrt{n}, \bar{\mu}_i + z_{0.45} sd_i/\sqrt{n})$, where sd_i is the sample standard deviation, and $\bar{\mu}_i$ is the sample mean of Y^{α_i} of the simulated samples. Using tables of the normal distribution, we obtain $|z_{0.45}| = |z_{0.55}| = 0.1256$. Using the statistical tool R, this involves invoking the commands *abs(qnorm(0.45))* or *abs(qnorm(0.55))*.

8.4.1 Reconstruction from data in intervals

Here we apply the result described in Section 8.2, namely the reconstruction from data in ranges. The results are displayed in Figure 8.1. In the left panel, we present the density from the data of the sample of size $N = 200$ and in the right panel, that of the data of size $N = 1000$. For comparison purposes, in both cases, we plot the lognormal density from which the data was sampled and the histograms of the sample for comparison.

Along with the curves, we plotted the histogram of the data to emphasize that the difference between the reconstructed densities and the histogram is not due to bad quality of the reconstruction. This assertion is borne by the discrepancy measures that we present below.

Again, to better visualize the reconstructions against the data, in Figure 8.2, we display the plots of the cumulative distribution function of the empirical data, the true distribution function, and the maxentropic reconstruction. In the left panel, we show the case of $N = 200$ and in the right panel that of $N = 1000$ data points.

Below we examine the discrepancies quantitatively. Let us also mention that once the maxentropic density is obtained, we can compute what the "true" (or implied) moments must be. These are listed in Table 8.3.

Let us repeat the results of the discrepancy analysis similar to that performed in Section 7.5. We consider again three quality reconstruction tests with two kinds of discrepancy measures and perform three pairwise comparisons shown in Tables 8.4 and 8.5. It is clear that the L_1 and L_2 discrepancies between the histogram and the true den-

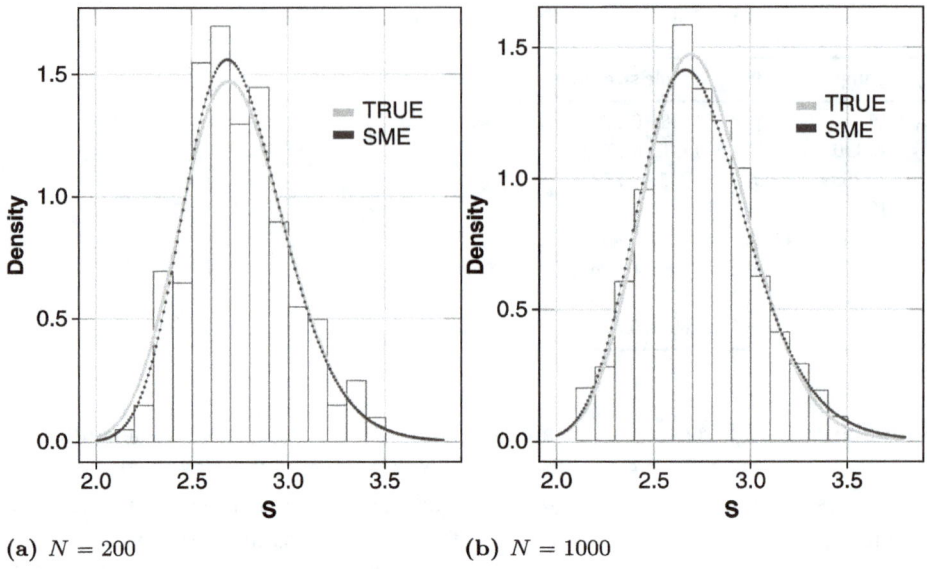

(a) $N = 200$ (b) $N = 1000$

Figure 8.1: Density distributions for different sample sizes.

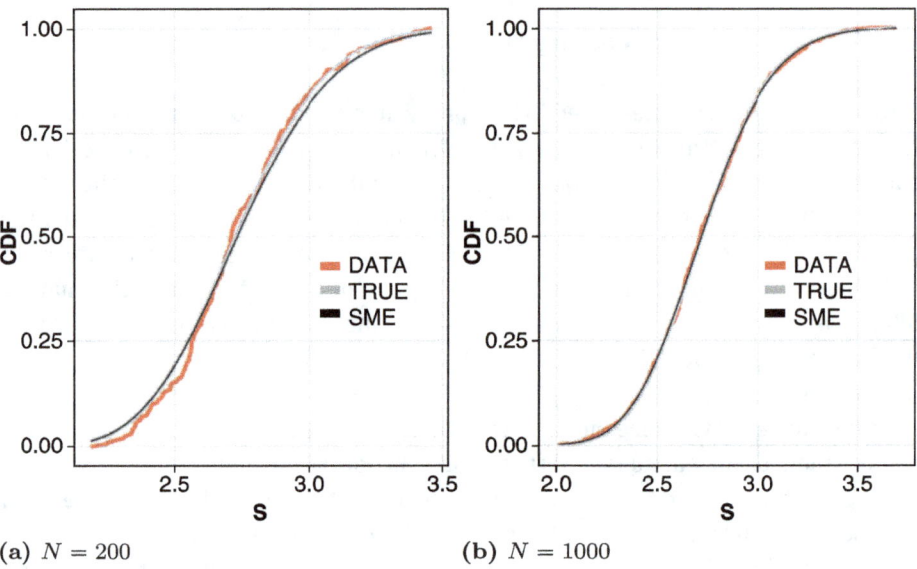

(a) $N = 200$ (b) $N = 1000$

Figure 8.2: Cumulative distribution functions for different sample sizes.

sity are quite similar to those for the histogram relative to the maxentropic densities. Also, the norm distances between the true density and the maxentropic densities are smaller than those between the histogram and density.

Table 8.3: Moments of the maxentropic densities.

Size	Moments of S							
	μ_1	μ_2	μ_3	μ_4	μ_5	μ_6	μ_7	μ_8
200	0.0176	0.1302	0.2559	0.3592	0.4405	0.5048	0.5565	0.5987
1000	0.0181	0.1319	0.2579	0.3612	0.4424	0.5066	0.5581	0.6002

Table 8.4: L_1 and L_2 distances between densities and histograms.

Approach	Hist. vs. true density		Hist. vs. maxent.		True density vs. maxent.	
	L1-norm	L2-norm	L1-norm	L2-norm	L1-norm	L2-norm
200	0.1599	0.1855	0.1504	0.1753	0.0527	0.0583
1000	0.1042	0.1077	0.1052	0.1158	0.0619	0.0577

Table 8.5: Discrepancy tests according MAE and RMSE.

Approach	Hist. vs. true density		Hist. vs. maxent.		True density vs. maxent.	
	MAE	RMSE	MAE	RMSE	MAE	RMSE
200	0.0158	0.0199	0.0104	0.0149	0.0106	0.0115
1000	0.0064	0.0076	0.0072	0.0090	0.0104	0.0119

The results in Table 8.5 confirm the visual analysis of Figure 8.2. This time, as the computations do not depend on the bin size, but rather on the data, the agreement of the pairwise comparisons is much better.

Notice as well that, as intuitively expected and as confirmed by the results displayed in Tables 8.4 and 8.5, the larger the size of the sample, the better the estimation results for both methods.

To conclude, we add that the size of the gradient of the dual entropy, which measures the quality of the reconstructions, was of the order 1×10^{-4} for both sample sizes. This value can be made much smaller by increasing the number of iterations.

8.4.2 Reconstruction with errors in the data

In this section, we consider the moments generated by the two samples as we did in Section 7.5, but we now assume that the difference between them and the true moments is due to a measurement error and solve the problem posed in (8.2). That is, instead of considering the data as given up to an interval, we search for the distribution that estimates true moments and a distribution in an auxiliary error range that provides us with an estimate of the error measurement. We will apply the results obtained in Section 8.2 without further ado. Let us recall the moments generated by the two samples considered in Section 7.5, which are listed in Table 8.6 for ease of access. Let us also mention that the intervals chosen for the error ranges were obtained by centering the confidence intervals obtained in (8.26),

$$(-\delta_i, \delta_i) = \left(-z^* \frac{sd_i}{\sqrt{n}}, z^* \frac{sd_i}{\sqrt{n}}\right). \tag{8.27}$$

Table 8.6: Moments of S for different sample sizes.

Size	Moments of S							
	μ_1	μ_2	μ_3	μ_4	μ_5	μ_6	μ_7	μ_8
200	0.0176	0.1304	0.2562	0.3596	0.4409	0.5051	0.5568	0.5990
1000	0.0181	0.1319	0.2579	0.3612	0.4424	0.5066	0.5581	0.6002

As already said, the application of the maxentropic procedure produces "true" moments, which we collect in Table 8.7.

Table 8.7: "True" moments provided by the maxentropic procedure.

Size	Moments of S							
	μ_1	μ_2	μ_3	μ_4	μ_5	μ_6	μ_7	μ_8
200	0.2409	0.4577	0.5820	0.6607	0.7146	0.7538	0.7835	0.8069
1000	0.2516	0.4640	0.5858	0.6631	0.7163	0.7550	0.7844	0.8076

The procedure provides us with an estimate of the measurement errors for each moment. These are listed in Table 8.8 for the two samples.

Table 8.8: Weights and estimated errors.

	k	1	2	3	4
$N = 200$	p_k	0.5184	0.4587	0.5066	0.5185
$N = 200$	ϵ_k	-5.11×10^{-5}	1.30×10^{-4}	-1.91×10^{-5}	-4.82×10^{-5}
$N = 1000$	p_k	0.5081	0.4793	0.5084	0.5088
$N = 1000$	ϵ_k	-1.13×10^{-5}	3.13×10^{-6}	-1.16×10^{-5}	-1.089×10^{-5}
	k	5	6	7	8
$N = 200$	p_k	0.5148	0.5055	0.4946	0.4837
$N = 200$	ϵ_k	-3.44×10^{-5}	-1.14×10^{-5}	1.03×10^{-5}	2.83×10^{-5}
$N = 1000$	p_k	0.5039	0.4996	0.4969	0.4955
$N = 1000$	ϵ_k	-4.30×10^{-6}	4.05×10^{-7}	2.84×10^{-6}	3.75×10^{-6}

Recall that the estimated error plus the reconstructed moment should add up to the observed moment. In our case the estimated errors turn out to be rather small, and both the reconstructed moments and the measured moments coincide.

Next, we display the true and the maxentropic densities along with the histogram in Figure 8.3.

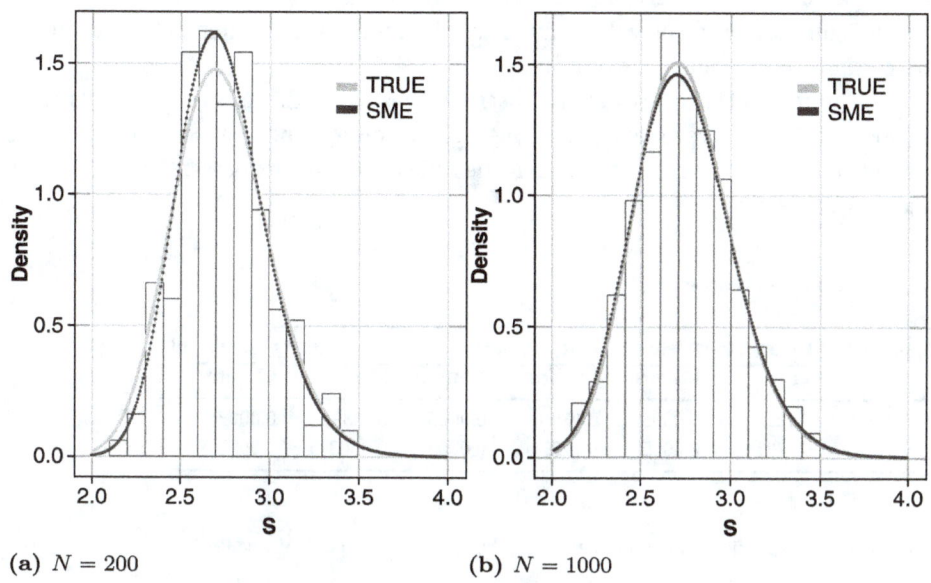

(a) $N = 200$ (b) $N = 1000$

Figure 8.3: Cumulative distribution functions for different sample sizes.

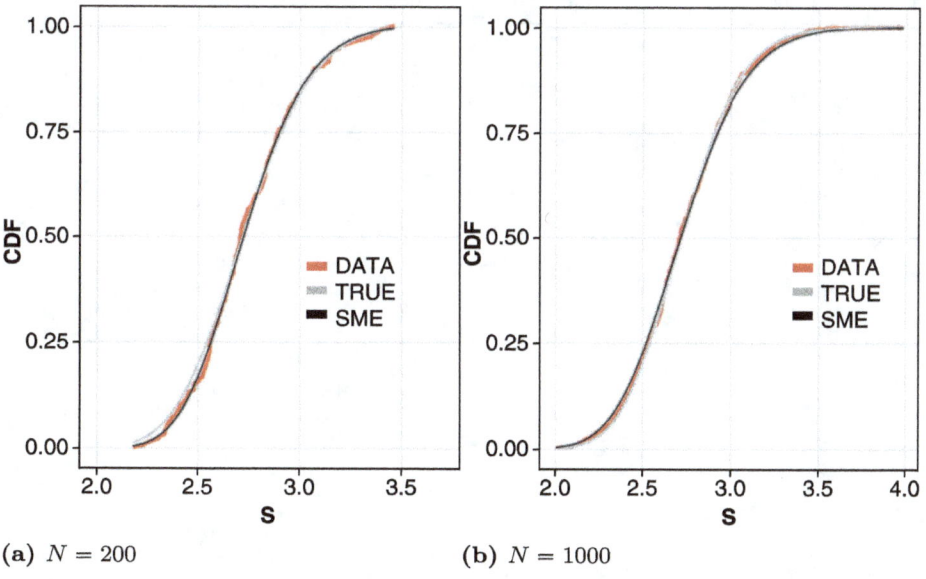

(a) $N = 200$ (b) $N = 1000$

Figure 8.4: Cumulative distribution functions for different sample sizes.

Again, the same comments that we made before apply, that is, the visual discrepancy between the densities and histogram is an artifact of the histogram generation process. This will be borne out by the numerical estimation of the distance between the densities as compared with the discrepancies between the cumulative distributions.

In Figure 8.4, we display the cumulative distributions. The statements made above are certainly borne out by looking at the plots.

The numerical measures of the discrepancies for both cases yield results similar to those of the previous cases. A measure of discrepancy between the densities among themselves and between the densities and histogram are presented in Tables 8.9 and 8.10, respectively.

Table 8.9: L_1 and L_2 distances.

Approach	Hist. vs. true density		Hist. vs. maxent.		True density vs. maxent.	
	L1-norm	L2-norm	L1-norm	L2-norm	L1-norm	L2-norm
200	0.1939	0.0005	0.1919	0.0005	0.1289	0.0003
1000	0.1703	0.0005	0.1663	0.0005	0.0803	0.0002

Table 8.10: Discrepancies among cumulative distributions according to MAE and RMSE.

Approach	Hist. vs. true density		Hist. vs. maxent.		True density vs. maxent.	
	MAE	RMSE	MAE	RMSE	MAE	RMSE
200	0.0158	0.0199	0.0151	0.0195	0.0211	0.0278
1000	0.0065	0.0077	0.0101	0.0119	0.0065	0.0075

9 Superresolution in maxentropic Laplace transform inversion

9.1 Introductory remarks

We have proposed the method of maximum entropy to determine a probability density of a positive random variable from the knowledge of its Laplace transform. In Chapter 4 we provided several examples of situations in which the problem appears, and in Chapters 7 and 8 we presented a collection of numerical examples in which we applied the method. The problem was stated as

$$\text{Find } f_S(s) \text{ such that } \quad \mathrm{E}[e^{-Sa_i}] = \int_0^\infty e^{-sa_i} f_S(s)\,ds = \mu_i; \quad i = 1,\ldots,K. \tag{9.1}$$

In Chapter 6 we mentioned that this problem can be solved by transforming it into a fractional moment problem on $[0,1]$, which can then be solved applying the maximum entropy technique. In the examples considered in Chapters 7 and 8, we saw that a small number of moments (eight to be precise) suffices to determine f_S with high accuracy. This fact was observed in [50] in the context of a comparison between maximum entropy based methods and other techniques for density reconstruction.

In our numerical work we have observed that in some situations the entropy of the reconstructed density with four moments does not change (actually does not decrease) very much when we considered eight moments. This happens to be relevant because the convergence in entropy and the convergence in the L_1 norm are related. The fact that the entropy stabilizes rapidly as the number of fractional moments increases may be responsible for the high accuracy in the reconstructions.

To make these statements more precise, let us denote by f_K the density reconstructed by means of the maxentropic procedure, determined by a collection of K moments. Below we shall see that as the number of moments increases, the difference in entropies between successive density reconstructions f_K becomes smaller and the f_K converge in L_1 to a limit density f_∞, which in our case it will be proven to coincide with the true density f_S. This argument was used in [37], in which the authors noted that successive integer moments lead to densities having entropies that decreased and became quite close. What we shall do below is make this argument precise.

This is rather curious and important. Observe that when using the maximum entropy method to invert Laplace transform we do not have to use complex interpolation to extend the data to an analytic function in the right half complex plane, but we only need to know the value of the transform at a small number of points to invert it. The fact that very few fractional moments provide a very good approximation to the true density is called *superresolution*.

The superresolution phenomenon has been noted in several applications of the maximum entropy method. To mention just a few references devoted to this matter using a related maxentropic procedure – the method of maximum entropy in the mean – consider for example [39], [63] and more recently [27]. The name superresolution is also used to describe the situation in which fine detail in a signal is obtained from a response in which much less detail is available. As examples of works along this line, consider [31], [17] and [16].

9.2 Properties of the maxentropic solution

So you do not have to flip pages back and forth, let us recall some properties of the solution to (9.1). Recall that after a change of variable $y = e^{-s}$ problem (9.1) becomes

$$\text{Find } f_Y(y) \text{ such that } \quad E[Y^{\alpha_i}] = \int_0^1 y^{\alpha_i} f_Y(y) dy = \mu_i; \quad i = 1, \ldots, K, \tag{9.2}$$

and once the $f_Y(y)$ is obtained, $f_S(s) = e^{-s} f_Y(e^{-s})$ provides us with the desired f_S. We also explained in Chapter 6 under what conditions fractional moments determined a solution to (9.2), and in Chapter 7 what the standard form of the solution to such a problem looks like. We saw that the maxentropic solution is given by

$$f_K(y) = \frac{e^{-\langle \lambda^*, y^\alpha \rangle}}{Z(\lambda^*)}. \tag{9.3}$$

As we are going to be working with different versions of f_Y, corresponding to different numbers of assigned moments, we drop the subscript Y and add the subscript K to emphasize that f_K solves (9.2) and satisfies the constraint given by the K moments. Here we denote by y^α the K-vector with components $y^{\alpha_i}: i = 1, \ldots, K$, and by $\langle \mathbf{a}, \mathbf{b} \rangle$ the standard Euclidean product of the two vectors. We also explained that λ^* is to be obtained by minimizing the (strictly) convex function

$$\Sigma(\lambda, \mu) = \ln Z(\lambda) + \langle \lambda, \mu \rangle, \tag{9.4}$$

overall $\lambda \in \mathbb{R}^K$. Not only that, if the minimum of $\Sigma(\lambda, \mu)$ is reached at λ^*, we then have

$$S(f_K) = -\int_0^1 f_K(y) \ln f_K(y) dy = \Sigma(\lambda^*, \mu) = \ln Z(\lambda^*) + \langle \lambda^*, \mu \rangle. \tag{9.5}$$

In Chapter 7 we introduced the notion of the relative entropy of a density f with respect to a density g, defined in the class of densities on $[0,1]$ by

For densities f and $g \in L_1([0,1])$ set $S(f,g) = -\int_0^1 f(y)\ln(f(y)/g(y))dy.$ (9.6)

From Chapter 7 we recall that the two $S(f,g)$ relevant for what comes below are contained in:

Lemma 9.1. *The relative entropy satisfies*
(1) $S(f,g) \leq 0$ *and* $S(f,g) = 0 \Leftrightarrow f = g$ *a.e.*
(2) $(1/2)\|f - g\|^2 \leq -S(f,g)$.

We mentioned there that first emerges from Jensen's inequality and the second is an exercise in [61].

9.3 The superresolution phenomenon

Let us begin by rephrasing the previous lemma as follows:

Lemma 9.2. *Let $M > K$ and let f_M and f_K be the maxentropic solution of the truncated moment problems* (9.2), *with M and K moments, respectively. Then*

$$S(f_M, f_K) = S(f_M) - S(f_K) \leq 0.$$
$$\|f_M - f_K\|^2 \leq -2S(f_M, f_K).$$

The second assertion is part of Lemma 9.1, and the first identity in the first assertion follows from (9.5). The inequality is backed by Lemma 9.1, but more importantly, it follows from the fact that both f_K and f_M share the first K moments and $S(f_K)$ has the largest entropy among such densities.

Let the fractional moments $\{\mu(\alpha_i) \mid i \geq 1\}$ be determined by a sequence $\{\alpha_i \mid i \geq 1\}$ satisfying the conditions of Lin's Theorem (Chapter 6). We now state a key assumption for the rest of the section:

Assumption A. *The density f with moments $\mu(\alpha_i)$ has finite entropy $S(f)$.*

This assumption is similar to the finiteness assumption in corollary 3.2 to theorem 3.1 in [12]. Before stating the result that we are after, let us comment on the intuition behind it and the practical application of the result. If the entropies on the truncated moment problems decrease to the entropy of the true density, we start with a problem with a few moments, say four, then find the maxentropic density and compute its entropy. Next, we increase the number of moments, one at a time say, and repeat the process. When the entropies of the successive densities change very little (one has to decide what that means), we stop and say that we got a satisfactory approximation at the density.

Of course, a strict mathematical minded critic may say that the decreasing sequence of entropies as the number of moments increases may have local clusters of very similar values, and then start to decrease again before the last cluster at the limit. Particularly if we are talking about small numbers (say eight moments) that criticism is true, but in the examples that we considered, eight moments provide quite a good approximation to the empirical density obtained from a very large number of data points.

The result that describes the superresolution property is contained in the following theorem:

Theorem 9.1. *Suppose that Assumption A holds true. Then, with the notations introduced above we have:*

1. $S(f_K)$ decreases to $S(f)$ as $K \to \infty$.
2. $\|f_K - f\| \to 0$ as $K \to \infty$.

Proof. Using an argument similar to the one used to prove Lemma 9.2, and from Assumption A, the following argument is clear. Since $S(f_K) \geq S(f)$, then the decreasing sequence $S(f_K)$ converges. Therefore, from the second assertion in (9.2) and the completeness of L_1, it follows that there is a function f_∞ such that $\|f_K - f_\infty\| \to 0$. That f_∞ integrates to 1 is clear, and by taking limits along a subsequence if need be, we conclude that $f_\infty \geq 0$, and therefore that f_∞ is a density.

Observe now that for any fixed K and the corresponding $a_i, 1 \leq i \leq K$, we have

$$\left| \mu(a_i) - \int_0^1 y^{a_i} f_\infty(y) dy \right| = \left| \int_0^1 y^{a_i} (f_M(y) - f_\infty(y)) dy \right| \leq \|f_M - f_\infty\| \to 0.$$

That is, the moments of f_∞ coincide with those of f, and according to Lin's Theorem in Chapter 6, we obtain that $f_\infty = f$, thus concluding the proof. □

Comment. That this result is similar to Theorem 3.1 in [12], and Assumption A is what allows us to make sure that the sequence of decreasing entropies $S(f_K)$ is actually a Cauchy sequence. This detail closes the gap in the argument in [37], and provides another approach to the problem considered in [63].

As far as the application of Theorem 9.1 to our numerical experiments goes, what we observed is that in going from four to eight decreasing fractional moments, the entropy of the reconstructed densities changed very little. Particularly if all we have to begin with is a histogram (of a dataset), and the fit of the maxentropic density to the histogram was quite good with four moments, and when you consider for eight moments the improvement is not that large, you may wonder whether there is something missing. The theorem says that no, there is nothing missing. This makes the application of the maxentropic procedure to the problem of reconstructing densities from Laplace transforms (converted into fractional moment problems) quite a convenient procedure compared to other numerical inversion techniques.

9.4 Numerical example

As a numerical example we consider a density reconstruction problem from numerical data. As with some of the previous examples, we consider a compound random variable $S = \sum_{k=1}^{\infty} X_k$, where N is a Poisson random variable of parameter $\ell = 1$, and the X_k are independent identically distributed random variables, all distributed according to a lognormal of parameters $m = 0$, $\sigma = 0.25$. Again, the lognormal model for the individual losses, or individual accumulated damages, was chosen because the Laplace transform of its density cannot be computed in closed form and one is forced to proceed numerically.

We considered a sample of size $M = 8000$ in order to compare the resulting density to the empirical density, and so that the numerical computation of the Laplace transform proceeds according to

$$\psi(\alpha_i) = \frac{1}{n} \sum_{k=1}^{M} e^{-\alpha_i S_k}; \quad i = 1, \ldots, M.$$

That this approximates $E[\exp(-\alpha S)]$ well is a consequence of the law of large numbers. Keep in mind that for each random value n of N, describing the number of losses during the time lapse, we have to aggregate a sample of n lognormals and sum them to obtain the corresponding total loss/damage. These total loss/damage measurements are the inputs for the computation of the Laplace transform.

There are a certain number of samples (time lapses) in which no shocks occur. This is taken into account by the fact that $P(N = 0) = \exp(-\ell > 0)$. This means that the distribution of total losses has a nonzero probability of being zero. To account for that before computing the density of the accumulated damage, from the Laplace transform we note that

$$\mu(\alpha) = E[e^{\alpha S} \mid N > 0] = \frac{\psi(\alpha) - e^{-\ell}}{1 - e^{-\ell}} = \int_0^{\infty} e^{-\alpha x} f_S(x) dx. \quad (9.7)$$

As fractional moments, or values of the transformed variable, we consider $\alpha_i = 1.5/i$ for $i = 1, \ldots 8$ when reconstructing from eight values of the Laplace transform. For the reconstruction from four values of α we considered every second one of these values, or the four even labeled ones if you prefer.

Once the $\mu_i \equiv \mu(\alpha_i)$ have been determined, we apply the maxentropic procedure to determine the density of the aggregate damage distribution. The results are presented in Figure 9.1, in which the densities obtained from four and eight moments are plotted along with the histogram.

In Table 9.1 we present the estimated reconstruction differences between the two reconstructions and the histogram. The L_1 and L_2 errors are computed in the obvious

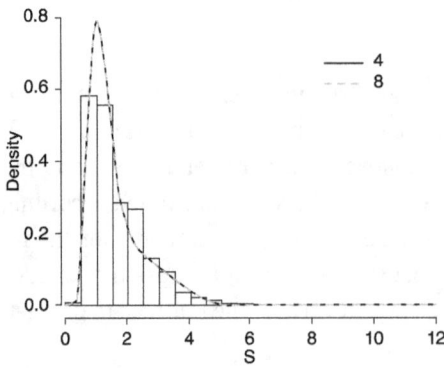

Figure 9.1: Density reconstructions and histogram.

Table 9.1: Errors and entropies for four and eight moments.

Error	Eight moments	Four moments
L_1-norm	0.2649	0.2653
L_2-norm	0.2099	0.2091
MAE	0.0216	0.02244
RMSE	0.0257	0.02671
Entropy	−0.567984	−0.565490

way, by discretizing the integral in such a way as to coincide with the bins of the histogram. The MAE (mean absolute error) and RMSE (root mean square error) are two common discrepancy measures between data and observations. The results suggest that the maximum entropy method applied to the empirically computed Laplace transform of the aggregate damage severity yields a rather good result with eight points, in case one feels that reconstructing from four data points may not be that good.

10 Sample data dependence

10.1 Preliminaries

In practice, the situation is closer to the examples considered in the previous chapters. That is, the situation in which all we know about the losses observed is that they can be modeled by compound random variables of the type $S = \sum_{n=1}^{N} X_n$, but where we do not necessarily have an explicit mathematical model for the frequency of losses N nor for the individual losses X_n. All that we have is empirical data about the total loss S during each observation period.

To be specific, the data that we have at the end of each observation period consists of a collection $\{n; x_1, \ldots, x_n\}$, where n is the number of risk events and each x_k denotes the loss occurring at the k-th risk event. The aggregate loss for that observation period is $s = \sum_{k=0}^{n} x_k$. When $n = 0$ there were no losses, the sum is empty and $s = 0$. When we need to specify the year j we shall write $(n_j, x_1, \ldots, x_{n_j})$ and $s_j = \sum_{k=0}^{n_j} x_k$. Suppose that the record consists of M years of data, therefore an observed sample (of losses) will be an (M) vector $\omega = (s_1, \ldots, s_M)$, in which each s_j is as just described.

We saw that the first thing to do to estimate the probability density of the losses is to compute the empirical Laplace transform by

$$\psi(\alpha) = \frac{1}{M} \sum_{j=1}^{M} e^{-\alpha s_j}. \tag{10.1}$$

Later on we shall consider the moments corresponding to K values of the parameter α. Since the distribution function of S has a probability $P(N = 0) = P(S = 0) > 0$ at $S = 0$, to determine the probability density of the losses we have to condition out this event and replace $\psi(\alpha)$ by

$$\mu(\alpha) = \frac{\psi(\alpha) - P(N = 0)}{1 - P(N = 0)}, \tag{10.2}$$

where $P(N = 0)$ is estimated as the fraction of the number of years of observation in which there were no losses. Recall that to transform the Laplace inversion problem into a fractional moment problem we use the change of variables $y = e^{-s}$, after which we can rewrite (10.1) as

$$\psi(\alpha) = \frac{1}{M} \sum_{j=1}^{M} y^{\alpha}, \tag{10.3}$$

which is the empirical version of

$$\psi(\alpha) = \int_0^1 y^{\alpha} dF_Y(y) = \int_0^{\infty} e^{-\alpha x} dF_S(x).$$

With this notation, we know that our problems consists of finding a density $f_Y(y)$ on $[0,1]$ such that

$$\int_0^1 y^\alpha f_Y(y)dy = \mu(\alpha)$$

in which the relationship between the $\psi(\alpha)$ and the $\mu(\alpha)$ is as detailed a few lines above. Once $f_Y(y)$ has been obtained, we know that the change of variables $f_S(x) = e^{-x}f_Y(e^x)$ provides us with the desired density.

As we have to emphasize the dependence of f_Y on the size M of the sample, we shall drop the Y and simply write f for it, and we shall use the notation $f_M(\omega, x)$ to denote the maxentropic density reconstructed from the collection of K moments given in (10.2).

We already saw that f_M can be obtained with the maxentropic method. What is interesting for us is that the maxentropic representation of the density obtained from the Laplace transform data can be used to study the sample dependence. The sample dependence is expected to be larger when the size of the sample is smaller. This is clear from (10.1). The convergence of the empirical Laplace transform to the true Laplace transform for each value of the parameter α is guaranteed by the law of large numbers, and the fluctuations of the empirical Laplace transform about its true value are described by the central limit theorem. Therefore, the reconstructed density f_M has to reflect the variability of the input somehow. The maxentropic representation of the solution will help us to understand how.

10.1.1 The maximum entropy inversion technique

Here we shall recall some results from Chapter 7 and Chapter 8, so that you, dear reader, do not have to flip pages back and forth. There we saw that the standard method of maximum entropy (SME) and the method of standard entropy with errors in the data (SMEE), provided us with a way to solve the following fractional moment problem:

$$\int_0^1 y^{\alpha_k} f(y)dy = \mu(\alpha_k) \quad \text{for } k = 0, 1, \ldots, K. \tag{10.4}$$

We set $\alpha_0 = 0$ and $\mu_0 = 1$ to take care of the natural normalization requirement on $f(y)$. We know that the solution to problem (10.4) provided by the SME method is given by

$$f^*(y) = \exp\left(-\sum_{k=0}^K \lambda_k^* y^{\alpha_k}\right), \tag{10.5}$$

which depends on the αs (or the $\mu(\alpha)$s) through the λs. It is customary to write $e^{-\lambda_0^*} = Z(\lambda^*)^{-1}$, where $\lambda = (\lambda_1^*, \ldots, \lambda_K^*)$ is a K-dimensional vector.

The generic form of the normalization factor is given by

$$Z(\lambda) = \int_0^1 e^{-\sum_{k=1}^K \lambda_k y^{a_k}} dy. \tag{10.6}$$

With this notation the generic form of the solution looks like

$$f^*(y) = \frac{1}{Z(\lambda^*)} e^{-\sum_{k=1}^K \lambda_k^* y^{a_k}} = e^{-\sum_{k=0}^K \lambda_k^* y^{a_k}}. \tag{10.7}$$

To finish, it remains to recall how the vector $\boldsymbol{\lambda}^*$ is to be determined. For that, one has to minimize the dual entropy

$$\Sigma(\boldsymbol{\lambda}, \boldsymbol{\mu}) = \ln Z(\boldsymbol{\lambda}) + \langle \boldsymbol{\lambda}, \boldsymbol{\mu}_Y \rangle \tag{10.8}$$

with respect to $\boldsymbol{\lambda}$ for each fixed $\boldsymbol{\mu}$. There $\langle \mathbf{a}, \mathbf{b} \rangle$ denotes the standard Euclidean scalar product and $\boldsymbol{\mu}$ is the K-vector with components μ_k, and obviously, the dependence on $\boldsymbol{a}$ is through $\boldsymbol{\mu}_Y$.

We mention that when the solution of this dual problem exists; then we have

$$H(f^*) := -\int_0^1 f^*(y) \ln f^*(y) dy = \Sigma(\boldsymbol{\lambda}^*, \boldsymbol{\mu}) = \ln Z(\boldsymbol{\lambda}^*) + \langle \boldsymbol{\lambda}^*, \boldsymbol{\mu}_Y \rangle. \tag{10.9}$$

There is an extended version of (10.4) that goes as follows. Determine f_Y and a K-vector $\boldsymbol{\eta}$ representing possible measurement errors by solving the following system:

$$\int_0^1 y^{a_k} f(y) dy + \eta_k = \mu(a_k) \quad \text{for } k = 0, 1, \ldots, K. \tag{10.10}$$

To simplify, we may suppose that measurement errors lie in a bounded interval $\eta_k \in [-c, c]$, which we take to be symmetric about 0. All these suppositions can be relaxed considerably at the expense of complicating the notation.

Under this assumption and consistent with the rest of the problem, we write $\eta_k = -p_k c + (1 - p_k) c$, that is as an expected value with respect to a measure that puts mass $(1 - p_k)$ at $-c$ and mass p_k at c. Now our problem consists of determining a density f_Y on $[0, 1]$ and parameters $0 < p_k < 1$ such that

$$\int_0^1 y^{a_k} f(y) dy - p_k c + (1 - p_k) c = \mu(a_k) \quad \text{for } k = 1, \ldots, K. \tag{10.11}$$

We already know from Chapter 8 that the maxentropic (SMEE) solution to this problem is given by

$$f^*(y) = \frac{e^{-\sum_{k=1}^{K} \lambda_k^* y^{a_k}}}{Z(\lambda^*)}$$

$$p_k = \frac{e^{c\lambda_k^*}}{e^{c\lambda_k^*} + e^{-c\lambda_k^*}}.$$
(10.12)

Here, the normalization factor $Z(\lambda)$ is as above. This time the vector λ^* of Lagrange multipliers is to be found by minimizing the dual entropy

$$\Sigma(\lambda, \mu) = \ln Z(\lambda) + \sum_{k=1}^{K} \ln(e^{c\lambda_k} + e^{-c\lambda_k}) + \langle \lambda, \mu \rangle.$$
(10.13)

As a side remark, notice that once λ^* is found, the estimator of the measurement error is, as is implicit in (10.11), given by

$$\eta_k = \frac{-ce^{c\lambda_k^*} + ce^{-c\lambda_k^*}}{e^{c\lambda_k^*} + e^{-c\lambda_k^*}}.$$
(10.14)

Notice that, although the formal expression for $f^*(y)$ is the same as that for the first method, the result is different because the λ^* is found by minimizing a different function.

10.1.2 The dependence of λ on μ

Since we are keeping $a_1, \ldots, a_K$ fixed, and the variability of the $\mu(a_k)$ will come from the variability of the sample used to estimate it, we are going to write μ_k for $\mu(a_k)$ from now on.

As the dependence of the reconstructed density on the sample μ comes in through λ, in order to study the sample variability we must begin establishing the dependence of λ on μ. To do that, we examine some properties of $\Sigma(\lambda, \mu)$ related to the minimization process. First, observe that in our setup, $\Sigma(\lambda, \mu)$ is a strictly convex and twice continuously differentiable function, defined on all $\mathbb{R}^K$. If we introduce

$$\Psi(\lambda) = \begin{cases} -\nabla_\lambda \ln Z(\lambda) \\ -\nabla_\lambda (\ln Z(\lambda) + \sum_{k=1}^{K} \ln(e^{c\lambda_k} + e^{-c\lambda_k})), \end{cases}$$
(10.15)

then the first order condition for λ^* to be a minimizer is that $\Psi(\lambda^*) = \mu$.

Notice that $\Psi(\lambda)$ is differentiable and its Jacobian (the negative of the Hessian matrix of $\ln Z(\lambda)$) is the negative of the covariance matrix of the fractional powers y^{a_i}, that is of the positive definite matrix with components $\mu(a_i + a_j) - \mu(a_i)\mu(a_j)$ with respect to the maximum entropy density.

Observe as well that μ has all of its components bounded by 1, thus the image of $[0, 1]$ by Ψ^{-1}, which is continuous due to the continuity and boundedness of the Hessian matrix, is compact. Thus the following assumption, which plays a role in the analysis carried out in the next section, is natural:

Assumption A. *Suppose that there is a ball $B(0,R)$ in $\mathbb{R}^K$ such that all solutions to $\Psi(\lambda^*) = \mu$ lie there for every $\mu \in [0,1]^K$. Suppose as well that the Hessian matrix*

$$\frac{\partial^2 \Sigma(\lambda^*, \mu)}{\partial \lambda_i \partial \lambda_j} = -\frac{\partial \Psi_i}{\partial \lambda_j}(\lambda^*)$$

has its upper eigenvalue uniformly bounded above and its minimum eigenvalues uniformly bounded away from 0 in $B(0,R)$.

The next result could have been stated in several places, but is related to the notation just introduced.

Lemma 10.1. *Let $f(y)$ be the density given by (10.7) (or (10.12)). Let $Z(\lambda)$ be as above, then the Hessian matrix of $\ln(Z)$ satisfies*

$$\nabla_\lambda \nabla_\lambda \ln(Z(\lambda)) = \mathbf{C}_f,$$

where $\mathbf{C}_f$ is the covariance matrix of y^a with respect to f, i.e.,

$$\mathbf{C}_f(i,j) = \int_0^1 y^{a_i+a_j} f(y)dy - \int_0^1 y^{a_i} f(y)dy \int_0^1 y^{a_j} f(y)dy$$

$$= \int_0^\infty e^{-(a_i+a_j)s} f(x)dx - \int_0^\infty e^{-a_j s} f(x)dx \int_0^\infty e^{-a_i s} f(x)dx,$$

where $f(y)dy = e^{-x} f(x)dx$ for the change of variables $y = e^{-x}$.

The proof is just a simple computation and it is left for the reader.

10.2 Variability of the reconstructions

The questions that we are interested in answering are:
(1) What is the limit of the f_M as $M \to \infty$?
(2) What is the mean of f_M?
(3) How does f_M fluctuate around its mean?

To establish the first result, denote by $\boldsymbol{\mu}_M(a)$ the vector of moments computed for a sample of size M, as in (10.2), and by $\mu_e(a)$ the exact moments. Denote as well by λ_M^* and λ_e^* the corresponding minimizers of (10.8). Denote as well by $f_M(x)$ and $f_e(x)$ the corresponding maxentropic densities.

Lemma 10.2. *With the notations just introduced, suppose that the risks observed during M consecutive years are independent of each other.*

Then $\mu_M(a) \to \mu_e(a)$ and $\lambda_M^* \to \lambda_e^*$, and therefore $f_M(x) \to f_e(x)$ when $M \to \infty$.

Proof. The first assertion emerges from an application of the law of large numbers to (10.1) and (10.2). Actually, the convergence is almost everywhere. The second assertion emerges from the continuity of Ψ^{-1}, which is a consequence of Assumption A. For the third assertion we invoke representations (10.7) and (10.12) as well as the first two assertions. □

To relate the sample variability of f_M to the sample variability of the $\mu(a)$, starting from $\lambda_M^* = \Psi^{-1}(\mu)$, and applying the chain rule, it follows that up to terms of $o(\delta\mu)$,

$$\delta\lambda_M^* = \mathbf{D}\delta\mu, \tag{10.16}$$

where $\mathbf{D}$ is the inverse matrix of the Jacobian of Ψ evaluated at λ_M^*. Actually, since the Jacobian of Ψ is the Hessian matrix appearing in the statement of Assumption A, then the Jacobian $\mathbf{D}$ of Ψ^{-1} is the inverse of the Hessian. Recall that the maxentropic solution to the inverse problem is

$$f_M(y) = \frac{1}{Z(\lambda_M^*)} e^{-\sum_{k=1}^{K} \lambda_k^* y^{a_k}} = e^{-\sum_{k=0}^{M} \lambda_k^* y^{a_k}},$$

in which a reference to the sample size M is made explicit. It is in the statement of the following results where the independence of the parameter c from the sample enters. Had we considered the intervals $[-c, c]$ to be sample dependent, the function Ψ defined in (10.15) would be sample dependent and the arguments that follow would not be true. With these notations another application of the chain rule leads to the proof of the following lemma:

Lemma 10.3. *With the notations introduced above, up to terms that are $o(\delta\mu)$, using the computation in (10.16),*

$$f_M(x) - f_e(x) = \sum_{i,j=1}^{K} (\mu(a_i) - e^{-xa_i}) f_e(x) D_{i,j} \delta\mu_j. \tag{10.17}$$

Here K is the number of moments used to determine the density and $\delta\mu_j = \mu_M(a_j) - \mu_e(a_j)$.

Let us carry out the necessary computations to verify the claim. To shorten the notation let us write y^a for the K-vector with components y^{a_i}, $i = 1, \ldots, K$, and $\delta\lambda = \lambda_M - \lambda_e$. Let us begin by rewriting (10.7) as

$$f_M(x) = \frac{e^{-\langle \lambda_M, y^a \rangle}}{Z(\lambda_M)} = \frac{e^{-\langle \lambda_e + \delta\lambda, y^a \rangle}}{Z(\lambda_e + \delta\lambda)} = \frac{e^{-\langle \lambda_e, y^a \rangle}}{Z(\lambda_e)} \frac{e^{-\langle \delta\lambda, y^a \rangle}}{Z(\lambda_e + \delta\lambda)/Z(\lambda_e)}.$$

Notice that the first factor in the last term on the right-hand side is the maxentropic solution $f_e(x)$ to the problem with exact moments μ_e. To continue, expand the numerator in the second factor up to the first order (or neglecting terms of second order) in $\delta\lambda$ as

$$e^{-\langle\delta\lambda,y^a\rangle} = 1 - \langle\delta\lambda,y^a\rangle.$$

To take care of the denominator, note that up to first order in $\delta\lambda$ we have

$$\frac{Z(\lambda_e^*\delta\lambda)}{Z(\lambda_e)} = 1 + \langle\nabla_\lambda \ln Z(\lambda^*), \delta\lambda\rangle.$$

Therefore, up to terms of first order in $\delta\lambda$ we have

$$f_M(x) = f_e(x)((1 + \langle\nabla_\lambda \ln Z(\lambda^*), \delta\lambda\rangle)(1 - \langle\delta\lambda, y^a\rangle)).$$

We know that for the exact density we have $\nabla_\lambda \ln Z(\lambda^*) = -\mu(a)$. Using this fact and invoking (10.16) we obtain the desired result.

The result in this lemma makes explicit the deviation of f_M from its exact value f_e up to first order in the $\delta\mu$. Notice that $\delta\mu$ is where all the randomness in the right-hand side lies, thus the result not only explains (or confirms) where the randomness comes from, but also quantifies it.

Observe that if we integrate both sides of (10.17) with respect to x we get 0 on both sides. Also, if we multiply both sides both e^{-xa_k} integrate to $\delta\mu_k$. This is due to the fact that $\mathbf{D} = -\mathbf{C}_{f^*}$ according to Lemma 10.1.

If we think of the f_M as random realizations of a density, they happen to be vectors in a convex set in $L_1(dx)$. The values $f_M(x)$ at each point can be thought as the components of those vectors, and results of the type of the central limit theorem for such values bear out in the simulations that we carried out.

However, these results are of not much use for direct applications because the exact moments $\mu_e(a)$ are unknown, nor is the maxentropic density f_e that they determine. But, a result that is potentially useful for the banking industry, goes along the following lines. Suppose that a bank has a relatively large number of branches, call it N_A, and that each of them has been collecting data for the last M years. If all the losses recoded by these agencies could be considered identically distributed, the computations in Lemma 10.2 lead to an interesting consequence. To establish it, we need to introduce extra labeling. Denote by $\mu_{M,m}(a_i)$ the i-th moment computed from the M-year data at branch $m = 1,\ldots,N_A$ of the bank. Denote as well by $f_{M,m}(x)$ the density reconstructed from the dataset at the branch m of the bank using a sample of size M. We then have:

Lemma 10.4. *Set*

$$\widehat{f}_M = \frac{1}{N_A}\sum_{m=1}^{N_A} f_{M,m}(x). \qquad (10.18)$$

Then, up to terms of $o(\delta\mu)$, we have $\widehat{f}_M = f_e(x)$.

Proof. Let us rewrite (10.17) in the proposed notation:

$$f_{M,m}(x) - f_e(x) = \sum_{i,j=1}^{K} (\mu_e(a_i) - e^{-x a_i}) f_e(x) D_{i,j}(\mu_{M,m}(a_j) - \mu_e(a_j)).$$

Summing over m we obtain

$$\hat{f}_M(x) - f_e(x) = \sum_{i,j=1}^{K} (\mu_e(a_i) - e^{-x a_i}) f_e(x) D_{i,j}(\mu_e(a_j) - \mu_e(a_j)) = 0$$

up to terms of order $o(\delta\lambda)$, which concludes the proof of this intuitive result. □

To measure of the variability we shall estimate the L_1 difference between the approximate reconstruction f_M for each sample and the true f_e density. For that we need the bound on the norm of the L_1 difference between densities by their relative entropy. Recall from Chapter 7 that the relative entropy of a density f with respect to a density g is defined by

$$f, g \in L_1([0,1]) \text{ set } \quad S(f,g) = -\int_0^1 f(y) \ln(f(y)/g(y)) dy. \tag{10.19}$$

In Chapter 7 we mentioned that some interesting properties of $S(f,g)$ relevant for what comes below are contained in the following result:

Lemma 10.5. *The relative entropy satisfies*
(1) $S(f,g) \leq 0$ and $S(f,g) = 0 \Leftrightarrow f = g$ almost everywhere.
(2) $(1/2)\|f - g\|^2 \leq S(f,g)$.

To estimate the relative entropy in terms of the entropies of the density we make use of (10.9) and Definition 10.19 to obtain

$$S(f_e, f_M) = \ln Z(\lambda_e^*) - \ln Z(\lambda_M^*) + \langle \lambda_e^*, \mu_e \rangle - \langle \lambda_M^*, \mu_e \rangle \cdot \langle \lambda_e^*, \mu_e \rangle. \tag{10.20}$$

Combining this with Lemma 10.2, we can easily verify the proof of the following assertion:

Proposition 10.1. *With the notations introduced above we have*

$$\|f_M - f_e\|_1 \leq (-4S(f_e, f_M))^{1/2} \to 0 \quad \text{as } M \to \infty. \tag{10.21}$$

Observe that a similar result would have been obtained had we considered $S(f_M, f_e)$. Of interest to us is that (10.21) allows us to estimate the sample size variability of the reconstruction for a given sample.

Actually, we can combine Lemma 10.2, equation (10.20) and Lemma 10.1 to further relate the deviations $\|f_M - f_e\|_1$ to the deviations of $\boldsymbol{\mu}_M$ from $\boldsymbol{\mu}_e$.

Proposition 10.2. *With the notations introduced above we have*

$$\|f_M - f_e\|_1 \leq (2\langle \delta\boldsymbol{\mu}, \mathbf{D}\delta\boldsymbol{\mu}\rangle)^{1/2} \to 0 \quad as\ M \to \infty. \tag{10.22}$$

Here $\delta\boldsymbol{\mu}$ and $\mathbf{D}$ are as above.

The proof is simple, and emerges from the fact that up to terms of order $o((\delta\boldsymbol{\mu})^2)$, $-S(f_e, f_M) = \langle \delta\boldsymbol{\mu}, \mathbf{D}\delta\boldsymbol{\mu}\rangle$.

10.2.1 Variability of expected values

Let us examine a few results that will be of use in Chapter 12 when we consider the application of the maxentropic densities to the computation of risk premiums. As expected values are to be computed using densities obtained from data samples, it is of interest to examine how sample variability is transferred from the sample to the expected values.

Let us begin with a simple consequence of Proposition 10.1 or Proposition 10.2.

Lemma 10.6. *Let H be a bounded, measurable function. Then, with the notations introduced above,*

$$E_{f_M}[H(S)] \to E_{f_e}[H(S)] \quad as\ M \to \infty.$$

The proof is easy. If K denotes any number such that $H(x) \leq K$, then

$$|E_{f_M}[H(S)] - E_{f_e}[H(S)]| \leq \int_0^\infty |H(x)|\,|f_M(x) - f_e(x)|dx \leq K\|f_M - f_e\|_1.$$

Now, invoke Proposition 10.1 or Proposition 10.2 to obtain the desired conclusion. For the next result we need the notation and conclusion in Lemma 10.3.

Lemma 10.7. *Let H be a bounded, measurable function. With the notations used in Lemma* 10.3 *we have*

$$E_{f_M}[H(S)] - E_{f_e}[H(S)] = \sum_{i,j} E_{f_e}[H(S)(\mu(\alpha_i) - e^{-\alpha_i S})]D_{i,j}\delta\mu_j.$$

The proof consists of substituting the claim in Lemma 10.3 in $E_{f_M}[H(S)]$. We shall examine the empirical behavior of $E_{f_M}[H(S)]$ in Chapter 12.

10.3 Numerical examples

The numerical example presented below is organized as follows:
(1) We describe how the sample is generated.
(2) We use the standard method of maximum entropy to determine a true density, which is to serve as reference for the sample variability analysis.
(3) Then we analyze numerically the sample variability with subsamples of the original sample.
(4) To conclude we analyze the effect of the sample variability on two possible ways to compute regulatory capital: value at risk (VaR) and tail value at risk (TVaR).

10.3.1 The sample generation process

To provide a touch of reality to the data, we consider a random variable resulting from summing eight compound variables, that is, the result of a double aggregation level. As models for the frequency of events we chose two Poisson random variables with parameters $\ell_1 = 80$ and $\ell_2 = 60$, four binomial frequencies of parameters ($n_3 = 70$, $p_3 = 0.5$), ($n_4 = 62, p_4 = 0.5$), ($n_5 = 50, p_5 = 0.5$) and ($n_6 = 76, p_6 = 0.5$). We considered as well two negative binomial frequencies with parameters ($n_7 = 80, p_7 = 0.3$) and ($n_8 = 90$, $p_8 = 0.8$).

To describe the individual risks, we denote by X_k the type of the individual loss compounded according to the k-th frequency in the previous list. The list goes as follows: X_1 and X_4 are chosen to be Champernowne densities with parameters ($a_1 = 20, M_1 = 85$, $c_1 = 15$) and ($a_4 = 10, M_4 = 125, c_4 = 45$). X_2 was chosen as a lognormal of parameters $(-0.01, 2)$. The variables X_3 and X_8 were chosen to have fat tails and to follow a Pareto distribution with (shape, scale) parameters given by $(10, 85)$ and $(5.5, 5550)$ respectively. Next in the list are X_5 and X_6, which were chosen to be gamma distributions with (shape, scale) parameters given respectively by $(4500, 15)$ and $(900, 35)$. Finally, X_7 was chosen to be a Weibull of type $(200, 50)$.

A moderately sized sample of $M = 5000$ was chosen. This is to insure a large enough sample to obtain a 'true' density to serve as the basis for comparison. To study the sample variability we chose different subsamples of sizes 10, 20, 50, 100, 500, 1000, and to numerically compute the Laplace transform we considered $K = 8$ moments determined by $\alpha_k = 1.5/k, k = 1, \ldots, 8$.

As before, since the size of the resulting aggregate losses has order of magnitude 10^4, and since the maxentropic methods use the Laplace transforms of the aggregate loss as starting point, to avoid numerical over(under)-flow, we scale the data prior to the application of the procedure, and reverse the scaling to plot the resulting densities. The simplest scaling consists of dividing the empirical losses by 10^4 as we did. Another possibility could be to introduce a scaled variable defined by $(s - \min(s))/(\max(s) - \min(s))$, where of course, $\max(s)$ and $\min(s)$ stand for the maximum and minimum values realized in the sample. Anyway, with all of this we are ready for the next stage.

10.3.2 The 'true' maxentropic density

Let us recall that by 'true' density we shall mean the density obtained from the large sample, which will be used for comparison. To obtain such density we apply the two maxentropic procedures mentioned above. We included the SMEE in the analysis to examine the possible effect of the numerical approximations involved in the computation of the fractional moments.

In Figure 10.1 we plot the maxentropic densities obtained applying the SME and the SMEE procedures to the full dataset. The histogram of dataset is plotted along for the sake of comparison. We have already mentioned that, since the dataset is very large, the maxentropic reconstructions will agree quite well with the empirical data.

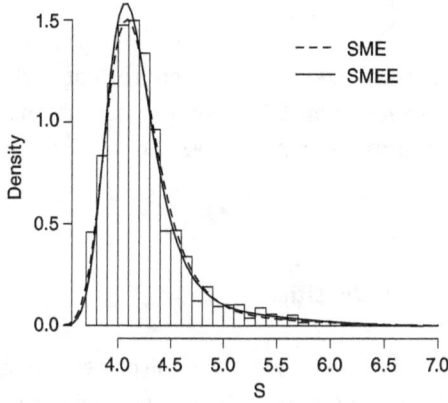

Figure 10.1: Histogram of the (scaled) total losses and maxentropic density.

Clearly the continuous densities are close enough to the histogram, and are closer to themselves. As the input data for both maxentropic techniques is the same, this means that the potential effect of the 'measurement' error is small. To quantify how close the curves are to the histogram, it is easier to use discrepancy measures like the mean absolute error (MAE) and the square root mean error (RMSE). Recall that these are computed as follows: If $\widehat{F}_s$ denotes the empirical distribution function and F_e denotes the 'exact' (reconstructed) distribution function, the measures of reconstruction error is to be estimated by

$$\text{MAE} = \frac{1}{N} \sum_{n=1}^{N} |\widehat{F}(x_n) - F_e(x_n)|$$

$$\text{RMSE} = \sqrt{\frac{1}{N} \sum_{n=1}^{N} (\widehat{F}(x_n) - F_e(x_n))^2},$$

where N is generic and stands for the size of the sample and the $\{x_n : n = 1, \ldots, N\}$ are the sample points. We do not compare the histogram to the continuous densities using standard L_1 and L_2, because the integration process will be bin dependent, resulting in a poorer estimate of the error.

In Table 10.1 we show the results of the computation of MAE and RSME to the full dataset and how the well the maxentropic densities fit the empirical data. Clearly, both maxentropic procedures yield quite good reconstructions.

Table 10.1: MAE and RMSE for a sample size of 5000.

Approach	MAE	RMSE
SMEE	0.005928	0.006836
SME	0.006395	0.009399

We also mention that the reconstruction error in the maxentropic procedure, that is the norm of the gradient of the dual entropy (8), was less than 10^{-6} in all reconstructions. This value is used as a criterion for stopping the iterations in the process of minimization of the dual entropy.

10.3.3 The sample dependence of the maxentropic densities

In this section we do several things. First, we examine the variability of the maxentropic density due to the variability of the sample. As we said above, the maxentropic techniques provide us with a density that has the given fractional moments. Thus, if the moments are not the true moments (as is the case with moments estimated from a small sample) the density may be quite different from the true density.

We begin by displaying the variability of the density reconstructed using the SME method.

The results displayed in the panels of Figure 10.2 should be understood as follows. To produce each panel of the figure, we applied the maxentropic procedures to 200 samples of the indicated sizes. For example, in the first panel the gray shadow consists of the plot of the 200 densities produced as outputs of the maxentropic procedure estimated from the moments produced by 200 samples of size $M = 10$. In each panel we also plot the average density $\hat{f}_M$ as well as the true density. Also, for each panel we average the moments and use them as input for the maximum entropy method, and we display the density obtained from such average moments along with the density obtained from the exact moments. That the last two coincide is no surprise according to the results in the previous section.

An important procedural matter is worth noting. Let us denote each of the 200 fractional moments of a sample of size M by $\mu_M^k(\alpha_i)$, where $i = 1,\ldots,8$ and $k = 1,\ldots,200$,

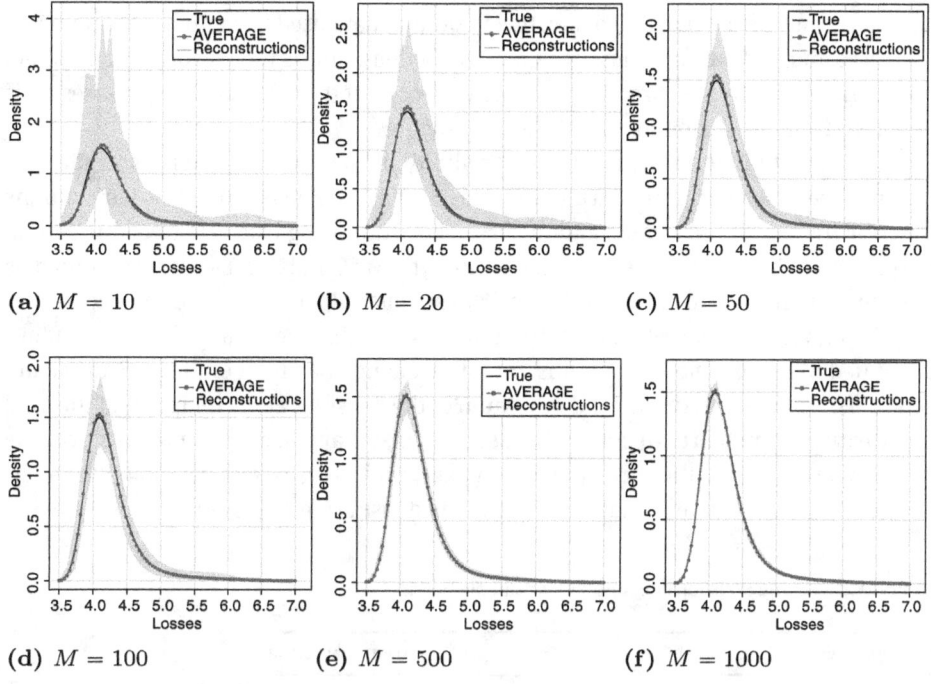

Figure 10.2: SME reconstructions with different sample sizes.

and let us denote by $f_M^k(x)$ the corresponding maxentropic density satisfying

$$\int_0^\infty e^{-\alpha_i s} f_M^k(x) dx = \mu_M^k(\alpha_i).$$

Note that from the linearity of that identity it follows that $\frac{1}{200} \sum f_M^k(x)$ is a density having moments $\frac{1}{200} \sum \mu_M^k(\alpha_i)$. But, even though the average density has the average moment (which may coincide with the true moment), it may not be the maxentropic density having that moment as a constraint, for the simple reason it may not maximize the entropy for that constraint. This is why we solve the maxentropic density for the average moment.

This side remark has importance in the case where we collect data from different but equivalent sources. Say different but equivalent branches of a bank have collected data independently for a short number of years. According to the results of the plots, if we estimate the same moments in each case and apply the maxentropic procedure to the average moments, we will obtain something quite close to the true density. This is in line with Lemma 10.2, for it means that we have a sample of a larger size to apply the maxentropic procedure (of size $200 \times M$) in our example.

As we move from panel to panel, the amount of data used to compute the moments increases and they become closer to their true values. This is the law of large numbers in action. Therefore, the spread in the indeterminacy of the true density decreases, as is apparent from the shrinking size of the gray shadow.

As a quantitative measure of the variability of the reconstructions, consider Table 10.2, the entries of which are to be understood as follows. The first two columns describe the size of the sample and the error measure being listed. To obtain the numbers listed in the next three columns, we listed the MAE and RSME errors computed as described in the previous section for the 200 samples of each size, and the percentiles mentioned were computed. In column number six we list the area of the gray shadow, computed in the obvious way. That is, for a fine enough partition of the horizontal (abscissa) axis find the maximum and minimum of the densities at the midpoint of the axis, and carry on from there. To obtain the last column, we averaged the moments over the 200 samples, obtained the corresponding maxentropic density, and computed the average discrepancies between that density and the densities in the sample.

Table 10.2: MAE and RMSE of SME for different sample sizes.

Size (M)	Error Meas.	1st Qu.	Mean	3rd Qu.	Area	Average
10	MAE	0.0500	0.0880	0.1150	2.625	0.0092
	RMSE	0.0590	0.1010	0.1310		0.0120
20	MAE	0.0359	0.0619	0.0811	1.523	0.0089
	RMSE	0.0400	0.0702	0.0915		0.0116
50	MAE	0.0203	0.0377	0.0510	0.9545	0.0082
	RMSE	0.0237	0.0429	0.0575		0.0106
100	MAE	0.0144	0.0266	0.0343	0.696	0.0053
	RMSE	0.0169	0.0304	0.0393		0.0066
200	MAE	0.0126	0.0194	0.0247	0.5375	0.0053
	RMSE	0.0148	0.0225	0.0285		0.0067
500	MAE	0.0081	0.0128	0.0163	0.3258	0.0055
	RMSE	0.0101	0.0153	0.0194		0.0076
1000	MAE	0.0067	0.0093	0.0108	0.2033	0.0054
	RMSE	0.0087	0.0115	0.0132		0.0078

To obtain the next figure, we repeat the same procedure as before, except that this time the densities were obtained using the SMEE method. This is to test whether specifying a measurement error improves the reconstructions. In the six panels of Figure 10.3 we did as described above.

Similar to the previous case, as we increased the amount of data the reconstructions improved, but as shown in Table 10.3 the improvement (relative to the results obtained using SME) as the sample size increase is small. The entries in Table 10.3 were produced and have the same meaning as those in Table 10.2.

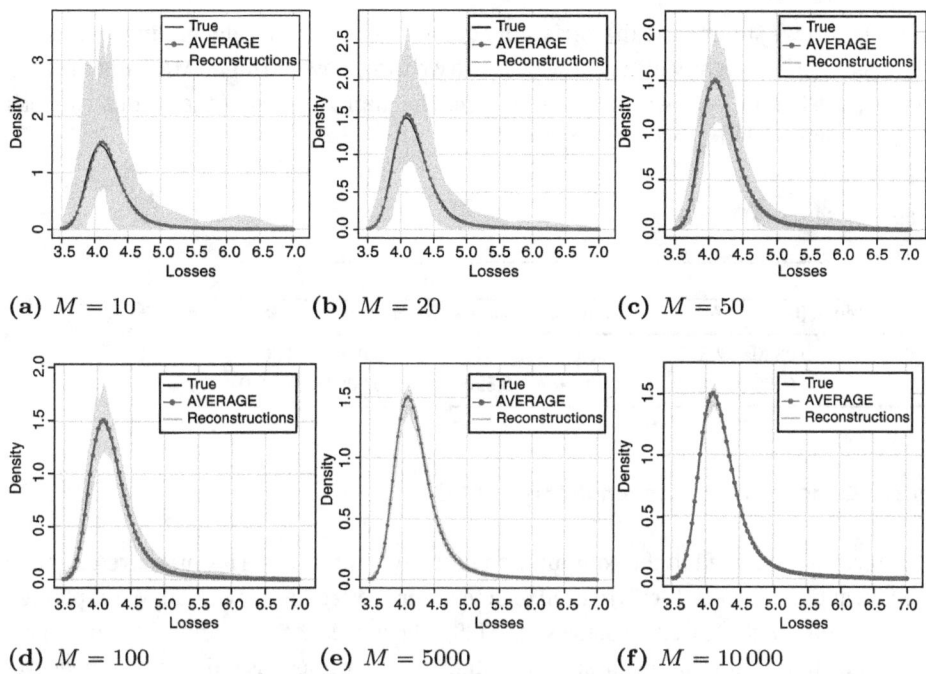

Figure 10.3: SMEE reconstructions with different sample sizes.

Table 10.3: MAE and RMSE of SMEE results for different sample sizes.

Size (M)	Error Meas.	1st Qu.	Mean	3rd Qu.	Area	Average
10	MAE	0.0419	0.0690	0.0898	2.619	0.0069
	RMSE	0.0514	0.0784	0.1030		0.0110
20	MAE	0.0360	0.0620	0.0820	1.759	0.0066
	RMSE	0.0420	0.0705	0.0918		0.0109
50	MAE	0.0198	0.0378	0.0500	1.044	0.0065
	RMSE	0.0240	0.0430	0.0582		0.0102
100	MAE	0.0142	0.0267	0.0353	0.690	0.0060
	RMSE	0.0168	0.0306	0.0398		0.0082
200	MAE	0.0125	0.0196	0.0247	0.552	0.0063
	RMSE	0.0147	0.0229	0.0270		0.0072
500	MAE	0.0083	0.0131	0.0165	0.294	0.0058
	RMSE	0.0101	0.0156	0.0199		0.0083
1000	MAE	0.0068	0.0093	0.0109	0.200	0.0057
	RMSE	0.0082	0.0114	0.0133		0.0082

In Table 10.4 we summarize the variability of the reconstructions depending on the size of the sample in a slightly different way. There we show how the mean MAE and the area of the gray shadow vary with the sample size. The improvement of the reconstructions as the sample size increases is apparent.

Table 10.4: Summary of results.

	10		100		1000	
	Mean (MAE)	Area	Mean (MAE)	Area	Mean (MAE)	Area
SME	0.0880	2.625	0.0266	0.696	0.0092	0.2033
SMEE	0.0690	2.619	0.0267	0.690	0.0093	0.2000

10.3.4 Computation of the regulatory capital

This section is devoted to the computation of the two most used risk measures, namely the VaR and the TVaR, which are used to determine the regulatory capital. We explained how to compute these risk measures as in [45]. The idea is that this analysis should provide us with insight into the possible variability of quantities essential in risk measurement.

In Table 10.5 we present a comparison of the values of the VaR and the TVaR computed from an empirical sample of size 5000, and the VaR and the TVaR computed using the SME and SMEE densities at the 95% and the 99% confidence levels. In the table, y stands for the confidence level.

Table 10.5: Comparison of VaR and TVaR at 95% and 99% for a unique sample of size 5000.

	y	Empirical	SME	SMEE
VaR	0.950	5.05	4.935	5.004
	0.990	5.72	5.755	5.772
TVaR	0.950	5.45	5.443	5.461
	0.990	6.05	6.0207	6.014

The sample used to build Table 10.5 might be considered large for operational risk purposes, and datasets corresponding to large disasters may comprise even smaller datasets. Therefore, computations like those leading to this table will have to take into account sample variability.

In Table 10.6 we consider two measures of variability of the VaR and the TVaR computed from the maxentropic densities obtained for 200 samples of the indicated sizes. In each cell we present the mean and the variance (within parentheses) of each risk measure, for each sample size. We do this at the 95% and the 99% confidence levels.

Table 10.6: Mean and standard deviation of the VaR and TVaR for 200 samples of different sizes.

Size (M)	VaR (95 %)		TVaR (95 %)		VaR (99 %)		TVaR (99 %)	
	SME	SMEE	SME	SMEE	SME	SMEE	SME	SMEE
10	4.96	4.87	5.30	5.156	4.331	5.283	4.328	5.634
	(0.4530)	(0.6740)	(0.5678)	(0.7984)	(1.083)	(0.8597)	(0.995)	(1.3138)
20	4.96	4.91	5.30	5.282	4.518	5.502	4.633	5.818
	(0.4536)	(0.5200)	(0.5678)	(0.6515)	(1.004)	(0.7537)	(1.004)	(0.8053)
50	4.97	4.925	5.39	5.386	5.003	5.688	5.162	6.058
	(0.2902)	(0.3254)	(0.382)	(0.4286)	(0.988)	(0.5463)	(1.069)	(0.5617)
100	4.96	4.931	5.44	5.43	5.457	5.779	5.694	6.016
	(0.1994)	(0.2258)	(0.251)	(0.2794)	(0.705)	(0.3626)	(0.768)	(0.3476)
200	4.97	5.013	5.45	5.46	5.624	5.766	5.871	6.016
	(0.1537)	(0.1828)	(0.1902)	(0.1995)	(0.395)	(0.2343)	(0.417)	(0.2498)
500	4.95	4.93	5.45	5.45	5.708	5.822	5.972	6.017
	(0.08963)	(0.1038)	(0.1136)	(0.1249)	(0.153)	(0.1539)	(0.159)	(0.1385)
1000	4.95	4.95	5.45	5.45	5.729	5.828	5.977	6.064
	(0.05815)	(0.07033)	(0.07088)	(0.07765)	(0.109)	(0.1065)	(0.107)	(0.09838)

To finish, consider Table 10.7, in which we compute the 'true' VaR and the 'true' TVaR of small samples. Recall that the 'true' density of a small sample of size M was obtained by averaging the densities of the 200 samples of size M, and was shown to become closer to the true density of the total loss as M increased. We see that the same happens to the VaR and the TVaR, as described in Table 10.7. Besides the relevance of the table for possible applications, we note that as the sample gets larger, the VaR and the TVaR become closer to their true values.

Table 10.7: VaR and TVaR for the average of the maxentropic densities for different sample sizes.

Size (M)	VaR (95 %)		TVaR (95 %)		VaR (99 %)		TVaR (99 %)	
	SME	SMEE	SME	SMEE	SME	SMEE	SME	SMEE
10	4.81	4.82	5.32	5.35	4.54	5.64	5.97	5.97
100	4.95	4.92	5.42	5.38	5.67	5.72	5.99	6.05
500	4.95	4.95	5.49	5.45	5.78	5.79	6.09	6.09
1000	5.01	4.95	5.50	5.45	5.78	5.79	6.05	6.05

11 Disentangling frequencies and decompounding losses

Even though today's data collection tools may allow for a high level of discrimination, it is still plausible that in many fields loss data comes in aggregate form.

For example, when recording losses due to mistyped transaction orders, the information about the person typing the order or the nature of the transmission error is not recorded. Or consider for example the losses due to fraud, where the total fraud events at some location or fraud events of some type are known, but they may have different causes. Or an insurance company may record all claims due to collisions but only keep the data about the car owner and not about the person(s) causing the accident.

It may be important for the purpose of risk management, risk control or risk mitigation, to be able to determine from the total risk recorded the sources of the risk events and the individual losses at each event.

To state the problem to solve more precisely, we are presented with a total loss severity S, and we know it may be the result of an aggregation like

$$S = \sum_{h=1}^{M} \sum_{n=1}^{N_h} X_{h,n}, \qquad (11.1)$$

where M is the (unknown) number of risk types. For each possible risk type, N_h is the (unknown) risk frequency, and $X_{h,n}$ denotes the (unknown) individual losses when the risk event of type h occurs.

In order to disaggregate the total loss we have to perform two tasks. The first consists of determining the number M of risk types and their statistical nature and then to determine their statistical nature. The second task is to determine the statistical nature of the individual losses.

We shall see that under certain circumstances these tasks can be satisfactorily solved. In particular, in the simple cases in which the number of risk types is low and the frequencies belong to the Panjer $(a, b, 0)$ class, quite a bit can be said. The problem of determining the nature of the individual losses is a bit more complicated, but something can be said as well.

This chapter is split into two sections, in each of which we shall address one of the two issues mentioned above.

11.1 Disentangling the frequencies

Certainly, if we are given the total number of risk events (or of any number of events) that took place during some time, even if we knew the sum of the numbers of events of different nature that occurred during that time, to determine how many events of each type took place seems a hopeless task. But it so happens that if the events can be modeled distributions of the Panjer $(a, b, 0)$ class, a reasonable answer can be obtained, especially when we know the number of different types that are being aggregated.

For ease of access, we shall review here some material about the $(a, b, 0)$ Panjer class first introduced in Chapter 2. This is a very interesting class of models, comprising four of the standard parametric distributions: Poisson, binomial, negative binomial and geometric. The formal definition goes as follows:

Definition 11.1. Let N be a random variable taking positive integer values, and write $p_k = P(N = k)$ for $k \in \mathbb{N}$. We shall say that N is in the class $(a, b, 0)$ if there exist constants a, b such that

$$p_k/p_{k-1} = a + b/k, \quad \text{for } k = 1, 2, 3, \ldots. \tag{11.2}$$

Comment. The 0 in the definition of the class refers to the fact that the recurrence relation requires $p(0)$ to be specified so that $\sum_{n\geq 0} p_n = 1$. We have already mentioned that this class plays an important role in the modeling process, since they are a key part of a recursive method for the computation of the full probability distribution of the compound random variable. The reader should consult [78] to learn all about these matters.

In Table 11.1 we display the relationship between the parameters a and b along with the parameters of the four families of probability distributions mentioned above.

Table 11.1: Relation between parameters (a, b) and the discrete family distributions.

Distribution	a	b	p_0
Poisson	0	λ	$e^{-\lambda}$
Binomial	$-\frac{p}{1-p}$	$(n+1)\frac{p}{1-p}$	$(1-p)^n$
Neg. binomial	$\frac{\beta}{1+\beta}$	$(r-1)\frac{\beta}{1+\beta}$	$(1+\beta)^{-r}$
Geometric	$\frac{\beta}{1+\beta}$	0	$(1+\beta)^{-r}$

The key observation for the disentangling process is that the recurrence relation (11.2) can be rewritten as

$$p(k)/p(k-1) = a + b/k. \tag{11.3}$$

Observe now that the ratio defined by $r(k) = p(k)/p(k-1)$ satisfies $kr(k) = ka + b$. Therefore, if the parameters a and b are known, the plot of $kr(k)$ versus k is a straight line, or to phrase it differently, that knowing the parameters and $p(0)$ we can use the recurrence to determine $p(k)$ for all ≥ 1.

That is that nice, but just as nice is the fact that this may be turned around. That is, we may combine the $(k, kr(k))$ plot with a linear regression procedure to infer the coefficients a and b, from which the probabilities p_k may be obtained.

But even nicer, this plot may suggest whether there is more than one regression present. Thus, with the problem of disentangling a frequency model, if we have any means of determining these regressions present in the collective plot we might be able to determine the mixture.

It is also clear from (11.1) that, except in the last case, if the values of a, b are known, the value p_0 is completely determined. For example, if the plot consists of a line with negative slope a, this suggests that the underlying probability distribution is binomial. From a we determine the probability p of success, and then use b to determine n.

In Chapter 14 we devote several sections to the various statistical techniques that can be used either to determine the straight lines in a mixture or the number of clusters in a population. The last may be used to determine the possible number of lines, and the former to determine the equations of the lines. Below we shall present only a few simple cases of the methodology, thus complementing the examples developed in Chapter 2.

11.1.1 Example: Mixtures of Poisson distributions

Let us now consider the following dataset corresponding to the (daily) frequency of errors by an unknown number of typists transcribing data. The details are shown in Table 11.2. The meaning of the columns is the following: In the first column we show the number of errors occurring and, in the second, the frequency of that event. In the third column we show the relative frequency (the empirical estimate of $p(k)$), and in the last column we show the result of computing $k\frac{n_k}{n_{k-1}}$. The data was obtained by simulating two Poisson distributions of parameters five and 16 respectively.

The routine carried out to produce Table 11.2 will be carried out in all examples treated below, except that we shall not display the tables anymore. This routine was used in the simple examples treated in Chapter 5, but it is here that the usefulness of modeling with the class $(a, b, 0)$ is essential. In Figure 11.1 we display the histogram (bar diagram) determined by that data as well as the Panjer plot for the data in Table 11.2.

Clearly, the underlying data is separated into two populations. It appears that there are two typists, one making errors with a Poisson intensity of five and the other with a Poisson intensity of 16.

In Figure 11.2 we can see how the EM algorithm separates the two groups of data points contained in the sample described in Table 11.2. To each of these groups we apply a goodness of fit test to verify whether they do belong to a Poisson distribution of parameters $\lambda_1 = 5$ and $\lambda_2 = 16$, which furthermore coincide with the sample mean of each group obtained by means of the maximum likelihood technique.

As a perhaps more realistic example, consider the case treated in [43]. The aim of the example is to show the work involved in determining both the number of groups in the mixture and their statistical nature.

Table 11.2: Number of errors made by two typists.

# of events (typos)	Frequency	Relative frequency (p_k)	$k \frac{n_k}{n_{k-1}}$
0	4	0.003481288	–
1	20	0.017406440	5.00
2	51	0.044386423	5.10
3	84	0.073107050	4.94
4	105	0.091383812	5.00
5	105	0.091383812	5.00
6	88	0.076588338	5.02
7	63	0.054830287	5.01
8	39	0.033942559	4.95
9	22	0.019147084	5.07
10	11	0.009573542	5.00
11	30	0.026109661	30.00
12	40	0.034812881	16.00
13	49	0.042645779	15.92
14	56	0.048738033	16.00
15	60	0.052219321	16.07
16	60	0.052219321	16.00
17	56	0.048738033	15.86
18	50	0.043516101	16.07
19	42	0.036553525	15.96
20	34	0.029590949	16.19
21	26	0.022628372	16.05
22	19	0.016536118	16.07
23	13	0.011314186	15.73
24	9	0.007832898	16.61
25	6	0.005221932	16.66
26	3	0.002610966	13.00
27	2	0.001740644	18.00
28	1	0.000870322	14.00
29	1	0.000870322	29.00
30	0	0.000000000	–
Total	1149	1	

For this case we suppose that the aggregate risk has two sources, the frequencies of each of them being a Poisson distribution with parameters $\ell_1 = 2$, $\ell_2 = 8$, and that the individual severities X_1, X_2 follow a common lognormal distributions, $X \sim \text{LogNormal}(-1, 0.25)$. All the variables are supposed independent. Besides this, we consider a sample of size 500 to compute $S = S_1 + S_2$, and all that we record, or consider as given data, is the total number of risk events and the total loss in each sample.

This is the first step in our methodology and we shall describe it in much detail. In the left panel of Figure 11.3 we show a histogram of the frequency of losses. It is clearly suggestive of the existence of more that one subpopulation. In order to determine it, we plot the Panjer lines in the right-hand panel of Figure 11.3.

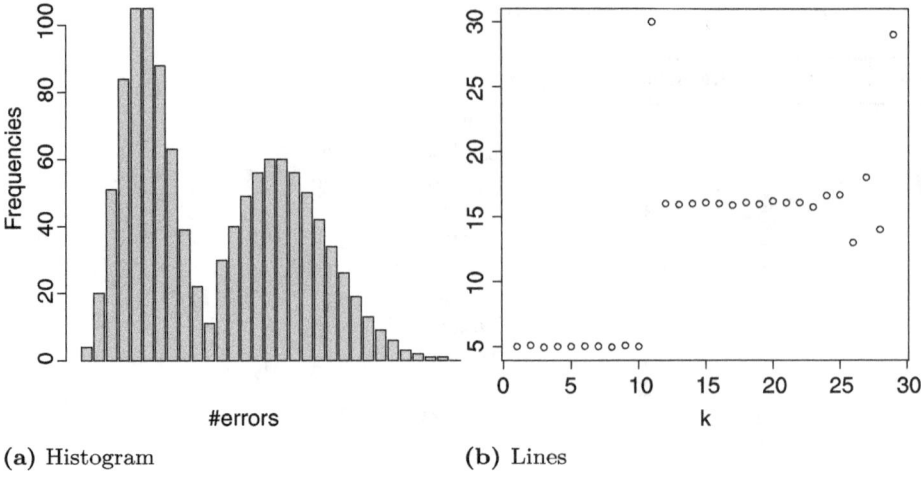

(a) Histogram (b) Lines

Figure 11.1: Histogram and Panjer plot for data in Table 11.2.

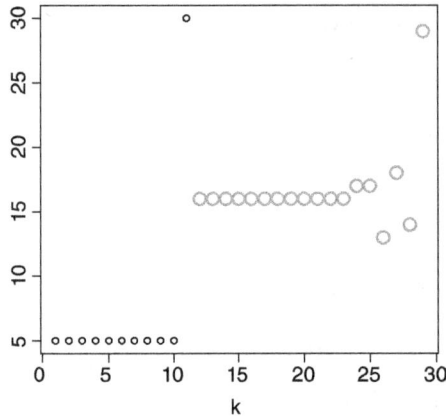

Figure 11.2: Result of applying the EM algorithm.

There we observe a group of values around the value two, and another group of points that is more dispersed. Also, towards the right end of the abscissa axis we observe some larger values. Additionally, the groups observed in Figure 11.3b show little slope, and the only significant increase in the values occurs at the end. This seems to indicate that the underlying distribution could be a mixture of Poisson distributions. By rescaling the vertical axis of Figure 11.3b, the slope of the points would look steeper, and a univariate negative binomial distribution can also be a possible candidate. But this possibility is not borne out by the shape of the histogram of Figure 11.3a.

To get a better idea of the number of groups that are present in the data sample, it is customary to utilize a variety of information criteria measures like the AIC, AIC3, BIC and ICL-BIC values (further detail about this is provided in Chapter 14, and a short explanation at the bottom of Table 11.3). These measures address the goodness of fit of

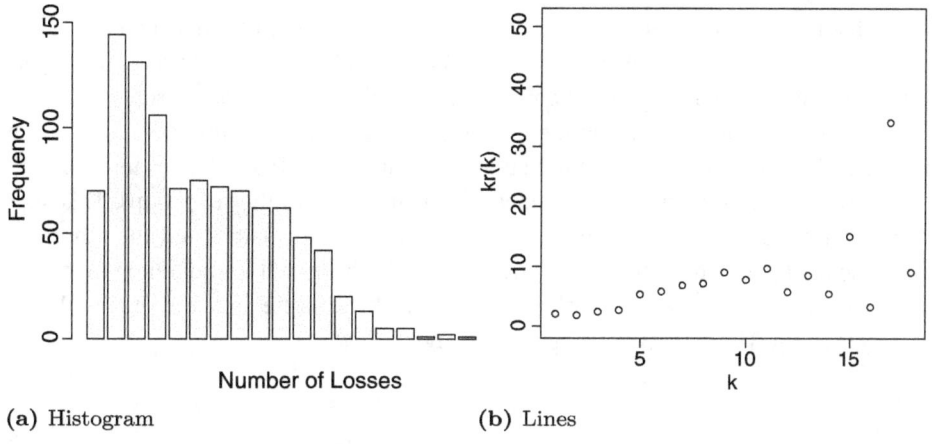

(a) Histogram (b) Lines

Figure 11.3: Histogram and Panjer plot example 2.

Table 11.3: AIC, AIC3, BIC, ICL-BIC & negentropy values (Case 1).

No. of components	AIC	AIC3	BIC	ICL-BIC	Negentropy
1	255.2	259.2	259.0	259.0	1
2	253.6	260.6	260.2	266.1	0.7763
3	231.5	241.5*	240.9*	244.9	0.9043
4	229.4	242.4	241.7	243.7*	0.9616*
5	228.4*	244.4	243.5	247.8	0.9292
6	229.8	248.8	247.7	251.4	0.9458

* Value of g given by criterion
AIC = $-2\,\text{loglik} + 2H$, AIC3 = $-2\,\text{loglik} + 3H$
BIC = $-2\,\text{loglik} + H\log(n)$, ICL = BIC + $2*$ ENT
Negentropy = $1 - \frac{\sum_{i=1}^{n}\sum_{l=1}^{H} -p_{il}\cdot\ln(p_{il})}{n\cdot\ln(H)}$
p_{il} is the posterior probability of the element i being in group l
H: number of groups, n: number of elements in the data

the clustering method (in our case, the EM method), and they are defined so that bigger is better. Additionally, we estimate the negentropy measure, which indicates how well discriminated or separated the classes seem to be, based on posterior probabilities (as before the higher the measure, the better). The results of these estimations are listed in Table 11.3, where we consider a number of subpopulations between one and six. This range was selected according to the result obtained with the sum of squared errors (SSE) for a broad number of clusters. This methodology is known in the clustering literature as the Elbow method (see Chapter 14), which consists of calculating the sum of the squared distance between each member of a cluster to its cluster centroid. This methodology also provides a possible number of clusters.

Table 11.3 shows that the information criterion is not very conclusive; these values indicate that the number of groups is between three and five. On the basis of the negentropy measure one may suppose that there are four subpopulations present.

Additionally, more advanced methodologies may be brought into the analysis. For example, the projection pursuit method of Peña and Prieto [80, 81] gives us a possible clustering without the need to introduce in advance the number of groups. This algorithm detects four groups in our data, two of them being clusters and the other two being isolated points or outliers (the largest values). This result is equivalent to the one obtained with the EM algorithm when the input is the discrete data and the number of groups is $H = 4$. The results are displayed in Figure 11.4a.

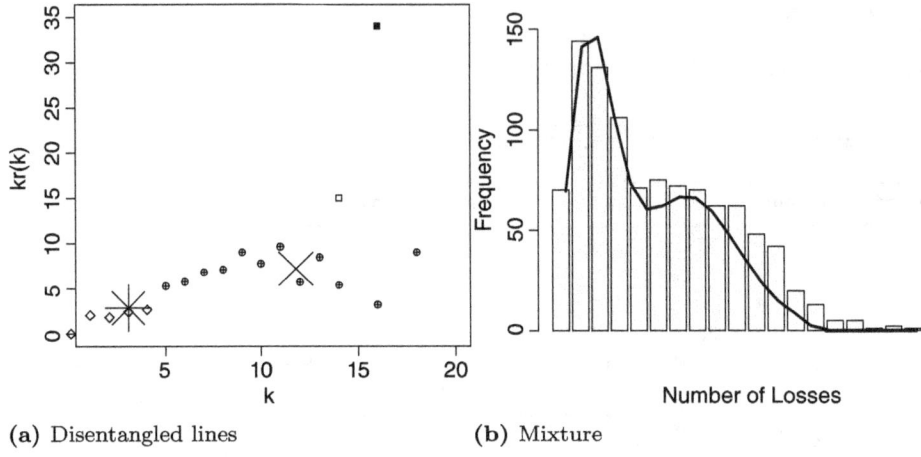

(a) Disentangled lines   (b) Mixture

Figure 11.4: Disentangled lines and detailed mixture.

There we see that the two largest clusters have almost zero slope, which indicates that the two groups follow a Poisson distribution. Combining the results with the k-mean, the Poisson parameters are the centers with values of 2.08 and 7.64 for each group. Rounding these values, the density obtained is shown in Figure 11.4b as a mixture of Poisson distributions in which the weights of the mixture are the proportions of points in each group provided by the disentangling procedure.

Looking at Figure 11.4b we conclude that this density mixture provides a satisfactory fit to the data. Considering the χ^2 goodness of fit test statistic we find a satisfactory fit ($\chi_o^2 = 9.512$, $p = 0.5747$, 11 degrees of freedom). Additionally, when comparing this result with a simple Poisson and a simple negative binomial model, the mixture model always gives superior fit.

11.1.2 A mixture of negative binomials

The negative binomial, which after the Poisson distribution is probably the most popular in operational risk, and perhaps the frequency models that use the negative binomial to model risk events in operational risk, will fit the available data better than the Poisson. This is due to the fact that it has two parameters, which provides more flexibility in shape; see for example [26], [74] or [35], where it has been used to explain the frequency of fraud. In particular, in this case it is also conceivable that different perpetrators take part in the process, which might show up in the presence of several modes in the frequency histogram.

Suppose for example that yearly fraud data frequency yields a histogram like that in Figure 11.5a, where one group has parameters $p_1 = 0.5$ and $r_1 = 10$, and the other one $p_2 = 0.4$ and $r_2 = 20$ (here $p = \beta/(\beta + 1)$).

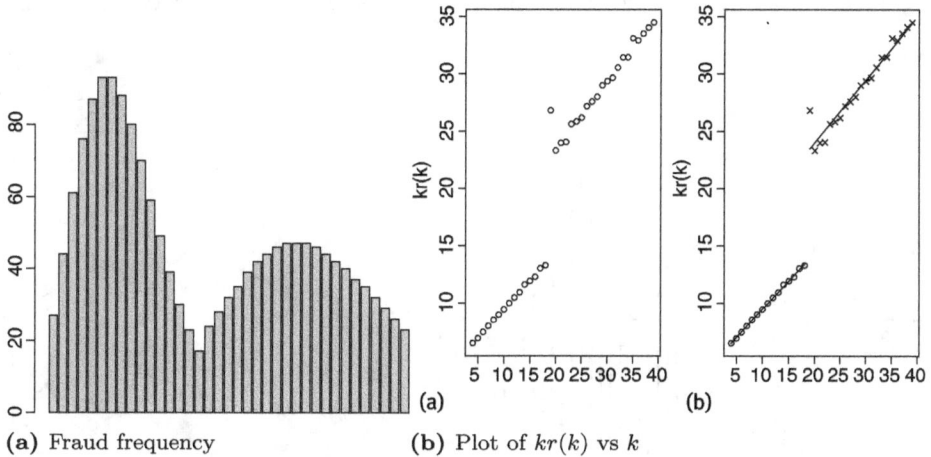

(a) Fraud frequency   (b) Plot of $kr(k)$ vs k

Figure 11.5: Disentangled lines and detailed mixture.

When the $(k, kr(k))$ plot is produced, we obtain a histogram like that in Figure 11.5b. The usefulness of the plot is again clear. It certainly suggests the presence of two different straight lines. The EM procedure tells them apart for us, yielding the plot in Figure 11.5b. The resulting parameters are $b_1 = 4.54(0.057519)$, $a_1 = 0.49(0.00486)$ and $b_2 = 13.10(0.94896)$, $a_2 = 0.54679(0.03203)$. Then, using the identities in Table 11.1, we obtain $\beta_1 = 0.9739$, $r_1 = 10.21639$, $p_1 = 0.506$ and $\beta_2 = 1.2064$, $r_2 = 24.96817$, $p_2 = 0.453$.

11.1.3 A more elaborate case

Let us now consider a case studied in [43], in which there are more sources of risk involved. The data we present below comes from a sample generated from the following four binomial distributions: $N_1 \sim$ binomial(15, 0.50), $N_2 \sim$ binomial(30, 0.65), $N_3 \sim$ binomial(30, 0.90) and $N_4 \sim$ binomial(40, 0.8). Somewhat more detailed information about the four frequencies is specified in Table 11.4.

Table 11.4: Binomial parameters of each group (original values).

-	μ	σ^2	p	n
Group$_1$	7.5	3.75	0.5	15
Group$_2$	19.5	6.825	0.65	30
Group$_3$	27	2.7	0.90	30
Group$_4$	32	6.4	0.8	40

These are listed for comparison with the results of the disentangling process. The histogram of the data and the plot of the Panjer relation appears in Figures 11.6a and 11.6b.

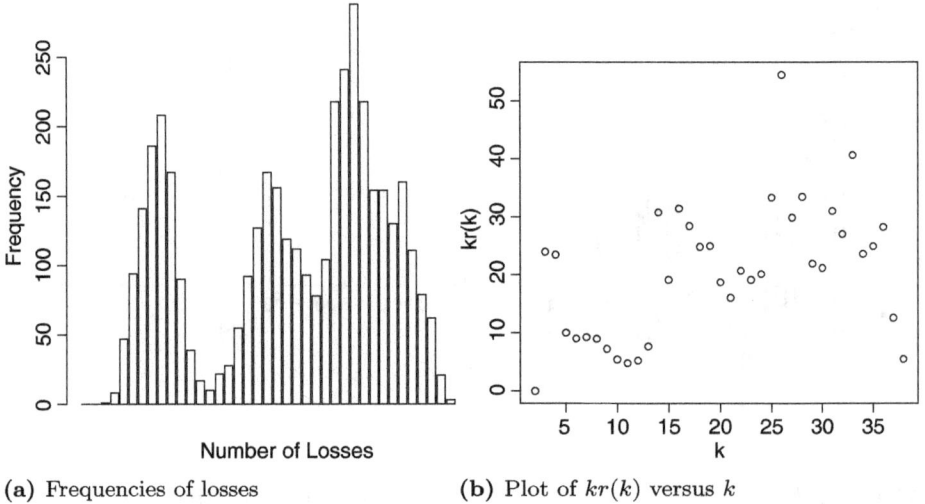

(a) Frequencies of losses (b) Plot of $kr(k)$ versus k

Figure 11.6: Histogram and plot of $kr(k)$ versus k.

To determine the number of subpopulations we applied the criteria summarized in Table 11.5. We settled for five groups according to the result suggested by the negentropy measure, as the other four criteria do not seem to be consistent. In the left panel of Figure 11.7a we observe the five groups obtained by the EM procedure. The first set of points is excluded and for the calculations the other four sets of data are used.

The parameters obtained by an application of the EM method are listed in Table 11.6 and should be compared with the original parameters listed in Table 11.4. The curve shown in Figure 11.7b is that of the mixed density, that is $P(N = k) = \sum_{j=1}^{4} P(N_j = k) f_j$, where f_j is the proportion of data points assigned to the j-th group by the disentangling procedure. We can say that the agreement is quite reasonable.

Table 11.5: AIC, AIC3, BIC, ICL-BIC and negentropy values (Case 5).

No. of components	AIC	AIC3	BIC	ICL-BIC	negentropy
1	573.8	577.8	580.4	580.4	1
2	564.2	572.2*	577.5*	586.0*	0.8330
3	561.1	573.1	581.1	593.6	0.8457
4	560.5	576.5	587.1	597.2	0.9011
5	557.8*	577.8	591.1	597.8	0.9434*
6	558.7	582.7	598.7	608.0	0.9291

* Value of g given by criterion
AIC = −2 loglik +2H, AIC3 = −2 loglik +3H
BIC = −2 loglik +H log(n), ICL = BIC + 2 ∗ ENT
Negentropy = $1 - \frac{\sum_{i=1}^{n}\sum_{l=1}^{H} -p_{il} \cdot \ln(p_{il})}{n \cdot \ln(H)}$
p_{il} is the posterior probability that the element i being in group l
H: number of groups, n: number of elements in the data

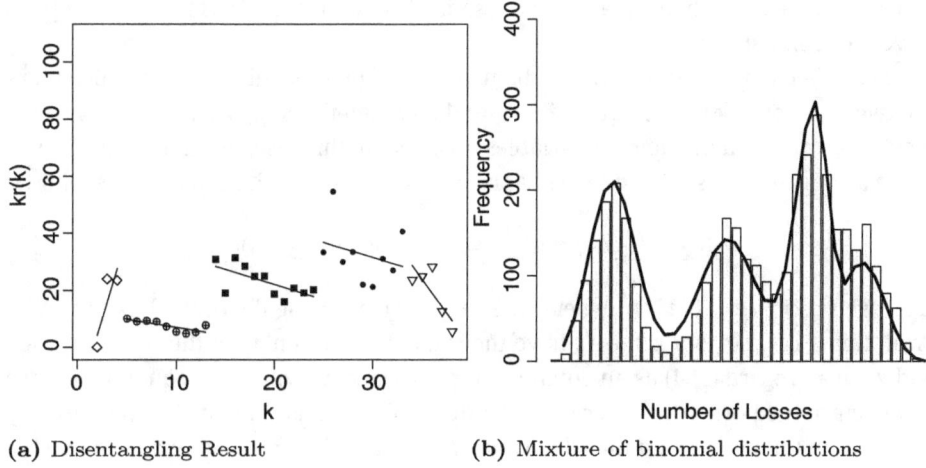

(a) Disentangling Result (b) Mixture of binomial distributions

Figure 11.7: Disentangled lines and distribution.

Table 11.6: Disentangled parameters for each group.

–	μ	σ^2	p	n
Group$_1$	7.67	4.94	0.36	22
Group$_2$	19.2	7.68	0.60	32
Group$_3$	27.3	2.457	0.91	30
Group$_4$	32.37	5.50	0.83	39

To sum up, results presented in these examples show that when we can model the frequency of losses by distributions in the Panjer $(a, b, 0)$ class, and when the number of risk sources is small, we can make use of various statistical techniques to determine the properties of the underlying distributions.

11.2 Decompounding the losses

In this section we address the remainder of the disaggregation problem, namely to determine the individual losses of each type. As that certainly turns out to be too much to ask for, we shall first examine some aspects of the problem and then present some examples.

Recall that what we are observing is samples from the distribution of

$$S = \sum_{h=1}^{M} \sum_{n=1}^{N_h} X_{h,n}.$$

Keeping in mind some of the examples presented in the previous section, clearly, unless the $X_{h,n}$ with the same value of h are identically distributed, that is, unless within each type of risk the individual losses have the same distribution, there is not much hope to solve the problem.

Let us begin by examining the problem in the simplest possible case, in which there are two types of risk with frequencies N_1 and N_2 and let us suppose that there is independence among all the random variables involved. In this case, the Laplace transform of the aggregate loss is related to the Laplace transforms of each partial loss as follows:

$$\psi_S(\alpha) = \psi_{S_1}(\alpha)\psi_{S_2}(\alpha) \quad \Leftrightarrow \quad G_{N_1}(\phi_1)(\alpha) G_{N_2}(\phi_2)(\alpha). \tag{11.4}$$

Here we set $\phi_h(\alpha) = \mathrm{E}[e^{-\alpha X_h}]$ where X_h is a generic variable distributed like the $X_{h,n}$. From this it is clear that if the X_h have the same distribution, then the $\phi_h(\alpha)$ coincide and we may regard (11.4) as an equation in $\phi(\alpha)$. Once we solve that equation and we have determined $\phi(\alpha)$, then we can invert the Laplace transform (at least numerically as shown in Chapter 7), and thus determine $f_X(s)$. This idea can readily be extended to the case in which there are more risk types as long as all the $\phi_h(\alpha)$ coincide.

But this is certainly too much to hope for. The question is: What is the next best that we can do? An interesting possibility presents itself when all the risk event frequencies are of the Poisson type.

To be specific, suppose that $N_i \sim \mathrm{Poisson}(\lambda_i)$, for $i = 1, \ldots, H$, and that the individual severities are distributed according to f_{X_i}. In this case, it is clear that (11.4) becomes

$$\psi(\alpha) = e^{-\sum_{h=1}^{H} \lambda_i (\mathrm{E}[e^{-\alpha X_h}] - 1)} = e^{-\lambda(\mathrm{E}[e^{-\alpha \tilde{X}}] - 1)}, \tag{11.5}$$

where we put $\lambda = \sum_{h=1}^{H} \lambda_h$ and manipulate the exponent as follows:

$$\sum_{h=1}^{H} \lambda_h \mathrm{E}[e^{-aX_h}] = \lambda\left(\sum_{h=1}^{H} \mathrm{E}[e^{-aX_h}]\frac{\lambda_h}{\lambda}\right). \tag{11.6}$$

Thus, in (11.5) $\widehat{X}$ is a random variable whose density is the mixture $f_{\widehat{X}} = \sum_{h=1}^{H} \frac{\lambda_h}{\lambda} f_{X_h}$. In other words, the aggregated risk is the result of compounding a risk produced with a Poisson intensity equal to the sum of the individual intensities and individual loss whose probability density equals that of a weighted average of the individual densities, where weights are given by the proportion of the individual frequency relative to the total frequency.

That idea was eventually extended by Wang in [96] into a result that we now state and prove.

Proposition 11.1. *Suppose that the H compound risks to be aggregated have Poisson frequencies N_i with a common mixing distributions $F(\theta)$ such that, given θ, the conditional distributions of the frequencies are given by $N_{|\theta} \sim P(\theta\lambda_i)$, and we suppose them to be independent. Let $S = \sum_{h=1}^{H} S_i$ be as above.*

Then $S \sim \sum_{n=1}^{N} \widehat{X}_n$ with $N_{|\theta} \sim P(\theta\lambda)$, where $\lambda = \sum_{h=1}^{H} \lambda_i$, and as above, the $\widehat{X}_n$ have common density $\sum_{h=1}^{H} \frac{\lambda_h}{\lambda} f_{X_h}$.

Proof. The proof is computational, and hinges on the fact that $\frac{\lambda_i}{\lambda} = \frac{\theta\lambda_i}{\theta\lambda}$ is independent of the mixing parameter θ. A simple computation like in (11.6) would take us from (11.5) to

$$\psi(a) = \int e^{-\theta\lambda(\mathrm{E}[e^{-a\widehat{X}}]-1)} dF(\theta) = \mathrm{E}\left[e^{-a\sum_{n=1}^{N} \widehat{X}_n}\right]$$

as claimed. □

Notice that for the proposed mixing we have $\lambda_i/\sum_j \lambda_j = \mathrm{E}[N_i]/\sum_j \mathrm{E}[N_j]$. The class of frequencies for which this is valid includes the Poisson and negative binomial random variables. The result contained in the proposition suggests what the best one can do regarding the decompounding problem is. That is, consider the total risk to be the compounding of individual risks that are distributed according to an equivalent individual loss defined as follows:

Definition 11.2. Let f_{X_h} denote the common probability density of the individual losses for the compound losses of the h-th type, $h = 1,\ldots,H$, and let N_h denote the respective loss frequency. The aggregated individual loss has a density given by

$$f_{\widehat{X}} = \sum_{h=1}^{H} \frac{\mathrm{E}[N_h]}{\mathrm{E}[N_{\mathrm{agg}}]} f_{X_h}, \tag{11.7}$$

which we shall refer to as the 'equivalent mixture'. Here we set $\mathrm{E}[N_{\mathrm{agg}}] = \sum_{h=1}^{H} \mathrm{E}[N_j]$ to be the mean of the aggregate loss frequency. That is, by $\widehat{X}$ we denote the individual loss, which, compounded according to N_{agg}, yields the total aggregate risk S.

Our goal is now more realistic and our original problem becomes first to determine the distributions of the N_i, and using that, to determine the distribution of the equivalent mixture of densities, from the knowledge of the aggregate risk data.

In the general case we are within the scope of the identifiability problem considered for example in chapter 3 of [95]. The importance of that work for us being that unless we know that the individual losses in the risk sources are identically distributed, from the aggregate loss we shall not be able to obtain the individual losses of the different risk types. At most, we shall be able to obtain the equivalent mixture of densities given (11.7). At this point it is interesting to note that the equivalent mixture coincides reasonably well with the distribution obtained by the decompounding procedure in the numerical experiments that we carried out.

We shall next examine a couple of examples considered in [44]. In each of these, part of the work consists of disentangling the frequencies. We shall now concentrate on the decompounding aspect of the problem.

11.2.1 Simple case: A mixture of two populations

We shall consider first the case of total losses produced by two sources. We shall suppose that the frequencies of each of them are a Poisson distribution with parameters $\ell_1 = 2$, $\ell_2 = 8$. We already saw above how to go about disentangling the two underlying frequencies.

Again, just to emphasize the power of the maxentropic procedure, we suppose that the individual severities X_1, X_2 follow a common lognormal distribution, $X \sim$ LogNormal$(-1, 0.25)$, which amounts to saying that their Laplace transforms can only be estimated numerically. All the variables are supposed independent. We consider a relatively small sample of size 500 to compute $S = S_1 + S_2$, and we only record the total number of risk events and the total loss in each sample.

Observe as well that since we are supposing that the individual losses in both cases are identically distributed, the equivalent mixture coincides with the individual loss density,

$$f_{\widehat{X}} = \frac{E(N_1)}{E(N_1) + E(N_2)} \times f_{X_1} + \frac{E(N_1)}{E(N_1) + E(N_2)} \times f_{X_2} \sim \text{LogNormal}(-1, 0.25).$$

To solve for the Laplace transform of the individual density when the frequency model is Poisson is easy now. We list in Tables 11.7–11.8 below the values of $\psi_S(\alpha)$ and those of $\phi_X(\alpha)$, obtained numerically as in Chapter 7 and using (11.4). Since we will apply both maxentropic methods, that is SME and SMEE, we also compute the confidence interval for the data about $\phi_X(\alpha)$.

The confidence interval for $\phi_X(\alpha)$ is determined by a bootstrapping procedure at the 5 % confidence level.

Table 11.7: Laplace transform of the aggregate losses and individual losses (Case 1).

k	1	2	3	4	5	6	7	8
$\psi(a_k)$	0.0064	0.0515	0.1229	0.1978	0.2671	0.3284	0.3818	0.4283
$\phi(a_k)$	0.4959	0.7034	0.7903	0.8380	0.8680	0.8886	0.9037	0.9152

Table 11.8: Intervals for the Laplace of the simulated individual losses (Case 1).

$\phi(a_k)$	Confidence Interval
$\phi(a_1)$	[0.4948, 0.4966]
$\phi(a_2)$	[0.7032, 0.7039]
$\phi(a_3)$	[0.7903, 0.7907]
$\phi(a_4)$	[0.8380, 0.8383]
$\phi(a_5)$	[0.8680, 0.8682]
$\phi(a_6)$	[0.8886, 0.8888]
$\phi(a_7)$	[0.9037, 0.9038]
$\phi(a_8)$	[0.9151, 0.9153]

Once we have reached this point, we apply the techniques proposed in Chapters 7 and 8 to obtain the maxentropic densities. The result obtained is plotted in Figure 11.8.

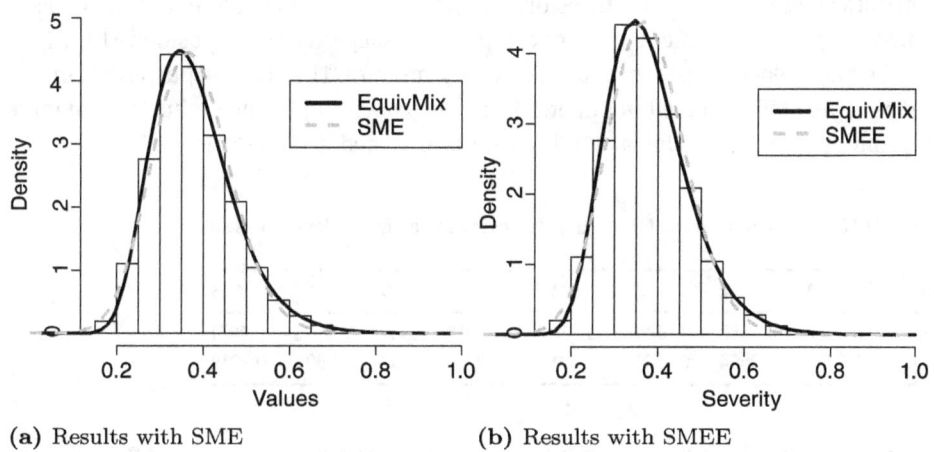

(a) Results with SME (b) Results with SMEE

Figure 11.8: Maxentropic individual densities.

Even though the description of the panels in Figure 11.8 is clear, let us emphasize that in the left panel we plot both the histogram of a sample of the true population as well as the reconstructed density. To measure the quality of the approximation we compute the

MAE and RMSE between the density and the empirical cumulative density. The results are listed in Table 11.9 displayed below.

Table 11.9: MAE and RMSE distances between reconstructed densities, original histogram and true densities.

Approach	Hist. vs. true density		Hist. vs. maxent.		True density vs. maxent.	
	MAE	RMSE	MAE	RMSE	MAE	RMSE
SME	0.0024	0.0029	0.0129	0.0143	0.0129	0.0143
SMEE	0.0024	0.0029	0.0143	0.0162	0.0142	0.0159

11.2.2 Case 2: Several loss frequencies with different individual loss distributions

Let us now consider a more elaborate example. The loss frequencies are those described and disentangled in the last example of the previous section. As individual loss densities generate the data, we consider the following four densities: $X_1 \sim \text{beta}(1, 25)$, $X_2 \sim \text{Weibull}(1, 0.1)$, $X_3 \sim \text{Frechet}(0.01, 2)$ and $X_4 \sim \text{gamma}(0.1, 3)$. They are associated in the same order with the binomials listed above in last example of the previous section.

This time we considered a sample of size 1000 and we recorded the total number of events and the total loss in each cycle of the simulation. In Table 11.10 we list the numerical values of the Laplace transform of the total loss $\psi_S(\alpha)$ and that of the Laplace transform of the equivalent mixture computed as suggested in Proposition 11.1, which is to be the Laplace transform of the equivalent mixture (11.7). For the purpose of implementing the SMEE detailed in Chapter 8, using a bootstrapping procedure we compute the confidence intervals for $\phi_{\widetilde{X}}(\alpha)$. The results are listed in Table 11.11.

Table 11.10: Laplace transform of the aggregate losses and individual losses (Case 2).

m	1	2	3	4	5	6	7	8
$\psi(a_k)$	0.0062	0.0688	0.1624	0.2524	0.3302	0.3957	0.4506	0.4971
$\phi(a_k)$	0.9425	0.9694	0.9791	0.9841	0.9872	0.9893	0.9908	0.9919

With this we are all set to apply the procedures presented in Chapters 7 and 8 once more. The result is plotted in the two panels of Figure 11.9; the left panel corresponding to the reconstruction using SME as in Chapter 7 and the right panel using SMEE as proposed in Chapter 8.

We only add that the histogram of the equivalent mixture is not available in a real situation. It is available for us to construct from the simulated data, and we compute it according to (11.7) for comparison purposes.

Table 11.11: Confidence intervals for the Laplace of the simulated individual losses (Case 2).

$\phi(a_k)$	Confid. interval
$\phi(a_1)$	[0.9425, 0.9426]
$\phi(a_2)$	[0.9694, 0.9695]
$\phi(a_3)$	[0.9791, 0.9792]
$\phi(a_4)$	[0.9841, 0.9842]
$\phi(a_5)$	[0.9871, 0.9872]
$\phi(a_6)$	[0.9892, 0.9893]
$\phi(a_7)$	[0.9908, 0.9909]
$\phi(a_8)$	[0.9919, 0.9920]

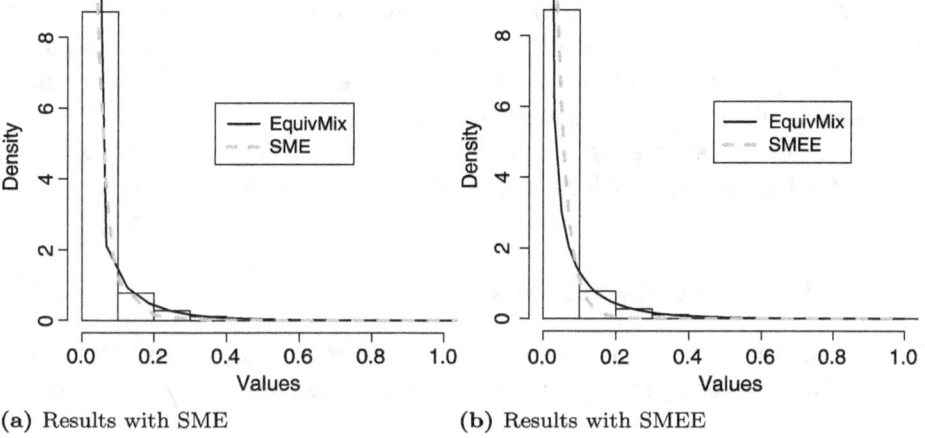

(a) Results with SME   (b) Results with SMEE

Figure 11.9: Maxentropic individual densities (Case 2).

To asses the quality of the reconstructions, in Table 11.12 we show the MAE and RMSE distances between the reconstructed densities, original histogram of the mixture and densities computed from the input data.

Table 11.12: MAE and RMSE distances between reconstructed densities, original histogram and densities.

Approach	Hist. vs. equiv. mix		Hist. vs. maxent.		Equiv. mix vs. maxent.	
	MAE	RMSE	MAE	RMSE	MAE	RMSE
SME	0.2120	0.2474	0.1270	0.1355	0.1052	0.1285
SMEE	0.2120	0.2474	0.1772	0.1854	0.0894	0.1009

Clearly, at least when the suspected number of risk types is small, the combination of the disentangling technique plus the maxentropic procedure yields an efficient approach to determining the frequencies of the risk events as well as that of an equivalent mixture of densities that plays the role of individual risk density for the aggregated loss.

12 Computations using the maxentropic density

In the previous chapters we devoted much effort to solving the problem of determining probability density from a few values of its Laplace transform, and to testing the quality of these reconstructions. In particular, we supposed that the unknown density is that of a positive compound random variable used to model losses. In this chapter we carry out some numerical computations with the maxentropic densities obtained in the previous chapters to answer some typical questions.

As one of the important applications of the methodology is to describe operational risk losses in a bank, or accumulated claims in an insurance company, the important applications to be considered include the VaR (value at risk) and the TVaR (tail value at risk), which are two of the standard measures required by a regulator to determine the capital to be set apart to cover potential losses. After that we shall consider some of the standard computations needed to compute the risk premium, that is, we shall compute the expected values of typical random variables.

But we shall once more begin with a theme treated in Chapter 5, related to the fact that in most loss models there is a positive probability that there are no losses.

12.1 Preliminary computations

Recall from Chapter 5 that in Lemma 5.1 we established that the density of losses was obtained by conditioning on the occurrence of losses according to

$$f_S(x) = \frac{d}{dx} P(S \leq x \mid N > 0). \tag{12.1}$$

When we want to compute expected values of quantities related to the total loss, we need to take that detail into account.

Let us begin with a generic result. The following is essentially a corollary of Lemma 5.1, but of more applicability:

Proposition 12.1. *Let $H: (0, \infty) \to \mathbb{R}$ be a bounded or positive measurable function. Then*

$$E[H(S)] = H(0)P(N = 0) + (1 - P(N = 0)) \int_0^\infty H(s) f_S(s) ds.$$

Proof. Since $\{N = 0\} = \{S = 0\}$, we have

$$E[H(S)] = E[H(S) \mid S = 0] P(S = 0) + E[H(S) \mid S > 0](1 - P(S = 0))$$
$$= H(0)P(N = 0) + (1 - P(N = 0))E[H(S) \mid S > 0]. \tag{12.2}$$

To conclude the proof, invoke (12.1). □

This means that computations made with the maxentropic densities have to be scaled when one wants to obtain unconditional expected values of quantities of interest.

In the remaining sections of the chapter we shall see how these remarks influence the computation of the quantiles of S from the knowledge of f_S.

12.1.1 Calculating quantiles of compound variables

Recall that if S denotes a continuous positive random variable, with density f_S, its q-quantile V_q is obtained by solving the equations

$$\int_{V_q}^{\infty} f_S(x)dx = 1 - q \quad \Leftrightarrow \quad \int_{0}^{V_q} f_S(x)dx = q$$

where, of course, $0 < q < 1$.

But when we deal with random variables like the compound variable S that describes total loss, we have to be a bit more careful. Let us now establish the relationship between the quantiles of S and the quantiles of $S|_{\{S>0\}}$, that is, the quantiles of the random variable with density f_S.

Note that from (12.1) or Proposition 12.1 it follows that the V_q quantile of S can be related to the V_{q^*}-quantile of f_S as follows. Note that if V_q is defined by $P(S \leq V_q) = q$, from Proposition 12.1 we know that

$$P(S \leq V_q) = P(S = 0) + (1 - P(S = 0)) \int_0^{V_q} f_S(s)ds = q \quad \Leftrightarrow \quad \int_0^{V_q} f_S(s)ds = \frac{q - P(S = 0)}{1 - P(S = 0)}.$$

Or, if we define

$$q^* = \frac{q - P(N = 0)}{1 - P(N = 0)}$$

and the V_{q^*}-quantile of f_S as the number such that

$$\int_{V_{q^*}}^{\infty} f_S(s)ds = \frac{q - P(N = 0)}{1 - P(N = 0)} = 1 - q^*,$$

we can state:

Proposition 12.2. *Let $q > P(N = 0)$, then the q-quantile V_q of S (or of $P(S \leq x)$) is the same as the q^*-quantile V_{q^*} of f_S, where $q^* = (q - P(N = 0))/(1 - P(N = 0)) < q$.*

Once more, the condition $q > P(N = 0)$ reflects the fact that for compound random variables describing accumulated loss or damage, the distribution function $P(S \leq s)$ has a jump of size $P(N = 0)$ at 0. Notice that even though $q^* < q$, this does not imply any relationship between V_q and V_{q^*} because they are quantiles of different (although related) distributions. Notice as well that since the maxentropic densities f_S are positive, we do not bother about right or left quantiles.

The result in the lemma has to be taken into account when using empirical data to determine the quantiles of S and the threshold probability, say $q = 0.999$, and when the probability of no loss is small, say $P(S = 0) = 10^{-4}$, then $q^* = 0.9989$. In the numerical examples below we shall compute the quantile of f_S from the data and from its maxentropic representation.

12.1.2 Calculating expected losses given that they are large

The object of this example is to point out how to use the loss density to compute the expected loss beyond a given threshold, that is, we want to compute $E[S \mid S > V]$ for some $V > 0$ using the density f_S. We shall verify that actually the natural calculations can be carried out. Observe that

$$E[S \mid S > V] = \frac{E[SI_{(V,\infty)}(V)]}{P(S > V)} = \frac{E[SI_{(V,\infty)}(V) \mid S > 0]P(S > 0)}{P(S > V \mid S > 0)P(S > 0)} = \frac{\int_V^\infty s f_S(s) ds}{\int_V^\infty f_S(s) ds}. \quad (12.3)$$

According to the comments in the previous section, when we use (12.3) we have to be careful when identifying V with a quantile. It is either the q-quantile of S (if $P(S > V_q) = 1 - q$) or the q^*-quantile of $S|_{\{S>0\}}$, with q and q^* related as above.

12.1.3 Computing the quantiles and tail expectations

Lets us now present a result taken from [84], which allows us to compute the quantile and the tail expected value in one single computation:

Proposition 12.3. *Let $f_S(s)$ be a strictly positive and continuous density on $[0, \infty)$, and consider for any $0 < \gamma < 1$ the function U defined on $(0, \infty)$ by*

$$U(t) = t + \frac{1}{1-\gamma} \int_t^\infty (s-t) f_S(s) ds.$$

Then, $U(t)$ is convex, twice continuously differentiable, achieves its minimum at the γ quantile V_γ of f_S and its minimum equals $E[S \mid S > V_\gamma]$.

The proof is simple and it involves taking two derivatives.

The quantity $E[S \mid S > V]$ is called the tail expected value of S. If the α-quantile V_α of f_S is called the *VaR* (value at risk) at (confidence) level α, then $E[S \mid S > V_\alpha]$ is called the *TVaR* (tail value at risk) of f_S.

Comment. For the computations described below, the upper limit of the integral was set equal to $10\max(S)$, that is, ten times the maximum value in the sample obtained for S. (That is, the possible range of t was chosen as $(0, 10\max(S))$.)

Besides this analytical procedure, we can use the standard method to determine VaR and TVaR from the empirical data. In order to compute the *empirical* VaR we consider the values of $S > 0$ ordered in increasing size ($s_1 \leq s_2 \leq s_n$), and then we estimate it as $\widehat{\text{VaR}}_\gamma(S) \approx x([N(\gamma)])$, where $[a]$ denotes the integer part of the real number a. The estimation of the TVaR is obtained from the same ordered list of values as

$$\widehat{\text{TVaR}}_\gamma = \frac{1}{N - [N\gamma] + 1} \sum_{j=[N(\gamma)]}^{N} s_j.$$

The confidence intervals for the VaR and TVaR were calculated by resampling without replacement with random subsamples between 50 % and 90 % of the total data size.

12.2 Computation of the VaR and TVaR risk measures

12.2.1 Simple case: A comparison test

In this section we shall compute the VaR and the TVaR of a known density (in our case a lognormal density of parameters $(1, 0.1)$) computed in three different ways: First from a sample of the distribution; second, using the true density and the procedure described in Proposition 12.3; and third, using the maxentropic density calculated using the procedures described in Chapters 7 and 8, using the empirical data as the input. Again, and to emphasize once more, remember that there is no analytical expression for the Laplace transform of the lognormal density but its exact density is known, so here we shall actually test the power of the maxentropic approach as well.

The details of this example were described in Chapters 7 and 8 in the context of the determination of the density. Recall that we considered two samples, one of size 200 and one of size 1000, and we carried out the reconstruction and a study of its quality supposing that the data was exact, or when it was given up to a range or that it was collected with measurement errors. Here we shall just determine the VaR and TVaR as mentioned above.

Exact data

This corresponds to the first numerical example treated in Chapter 7. Below we present four tables, one for each risk measure and each sample size, in which VaR and TVaR are computed as specified above.

In the left column of Tables 12.1 and 12.2 we display the confidence levels at which VaR and TVaR were computed. In the next three columns we list, respectively, the empirical quantity as well as its confidence interval at the 95 % confidence. In the last two columns we list, respectively, the VaR and the TVaR computed with the true density and with the maxentropic density.

Table 12.1: VaR: Sample size 200.

y	Empirical	VaR$_{inf}$	VaR$_{sup}$	True density	SME
0.950	3.1884282	3.1473332	3.2437457	3.1986397	3.1926385
0.960	3.2437457	3.1857360	3.3136158	3.2346469	3.2286457
0.970	3.3136158	3.2162549	3.3491837	3.2766553	3.2706541
0.975	3.3451934	3.2437457	3.3779235	3.3006601	3.2946589
0.990	3.3892554	3.3491837	3.4239602	3.4266853	3.4146829
0.995	3.4561725	3.3892554	3.4561725	3.5107021	3.4926985
0.999	3.4561725	3.4239602	3.4561725	3.6967393	3.6607321

Table 12.2: TVaR: Sample size 200.

y	Empirical	TVaR$_{inf}$	TVaR$_{sup}$	True density	SME
0.950	3.3268312	3.2425054	3.3891441	3.3379850	3.3277078
0.960	3.3571341	3.2627653	3.4073819	3.3684512	3.3572957
0.970	3.3840441	3.2905055	3.4257257	3.4066716	3.3941155
0.975	3.3941053	3.3084260	3.4352135	3.4303371	3.4167206
0.990	3.4342822	3.3653890	3.4561725	3.5437877	3.5230580
0.995	3.4481195	3.3835894	3.4561725	3.6246062	3.5965116
0.999	3.4561725	3.3892554	3.4561725	3.8002637	3.7488995

We add that in the two tables mentioned above and in the next tables, the empirical confidence intervals are determined by resampling from the datasets, with samples of size up to 90 % of the original sample size.

Next we present Tables 12.3 and 12.4, in which everything is as above, except that the sample size used is 1000.

Risk measures when the data is given up to a range

Let us now present the tables containing the results of the computation of the VaR and the TVaR when the data is given in ranges. Numerical examples of density reconstruction in this case were given in Section 8.4.1. The results presented in Tables 12.5 and 12.6 fare or a sample size of 200.

The description of the entries of these two tables is as for the previous examples. To examine the effect of the sample size in the determination of the risk measures, we

Table 12.3: VaR: Sample size 1000.

γ	Empirical	VaR$_{inf}$	VaR$_{sup}$	True density	SME
0.950	3.2360580	3.1938743	3.2458468	3.2006401	3.2086417
0.960	3.2473619	3.2283227	3.2856532	3.2326465	3.2486497
0.970	3.3106142	3.2473619	3.3283750	3.2726545	3.2886577
0.975	3.3220161	3.2770023	3.3754777	3.3046609	3.3206641
0.990	3.4239602	3.3837184	3.4385691	3.4246849	3.4406881
0.995	3.4717304	3.4239602	3.5427976	3.5127025	3.5367073
0.999	3.6898073	3.4717304	3.6898073	3.6967393	3.7287457

Table 12.4: TVaR: Sample size 1000.

γ	Empirical	TVaR$_{inf}$	TVaR$_{sup}$	True density	SME
0.950	3.3547538	3.2968652	3.4185504	3.3379840	3.3553835
0.960	3.3821343	3.3181586	3.4559391	3.3684504	3.3868917
0.970	3.4191905	3.3443793	3.4966896	3.4067017	3.4265035
0.975	3.4394590	3.3600881	3.5232593	3.4303656	3.4510622
0.990	3.5373198	3.4138775	3.6834803	3.5437671	3.5689433
0.995	3.6208459	3.4291698	3.8357996	3.6246247	3.6531619
0.999	3.8343980	3.4471884	3.9789886	3.8002637	3.8370929

Table 12.5: VaR: Sample size 200.

γ	Empirical	VaR$_{inf}$	VaR$_{sup}$	True density	SME
0.950	3.1884282	3.1884282	3.2162549	3.1986397	3.1986397
0.960	3.2437457	3.2437457	3.2770023	3.2346469	3.2346469
0.970	3.3136158	3.3136158	3.3451934	3.2766553	3.2826565
0.975	3.3451934	3.3451934	3.3491837	3.3006601	3.3126625
0.990	3.3892554	3.3892554	3.4239602	3.4266853	3.4446889
0.995	3.4239602	3.4239602	3.4561725	3.5107021	3.5407081
0.999	3.4561725	3.4561725	3.4561725	3.6967393	3.7567514

Table 12.6: TVaR: Sample size 200.

γ	Empirical	TVaR$_{inf}$	TVaR$_{sup}$	True density	SME
0.950	3.3386299	3.3121873	3.3574699	3.3379850	3.3515103
0.960	3.3659274	3.3412500	3.3817269	3.3684512	3.3846741
0.970	3.3897737	3.3725905	3.4010767	3.4066716	3.4266536
0.975	3.3988697	3.3846290	3.4118711	3.4303371	3.4529036
0.990	3.4390701	3.4211680	3.4473142	3.5437877	3.5810123
0.995	3.4549336	3.4400664	3.4561725	3.6246062	3.6749027
0.999	3.4561725	3.4561725	3.4561725	3.8002637	3.8869329

repeated the computation for a sample size of 1000. That there is an improvement is clear from the entries of Tables 12.7 and 12.8.

Table 12.7: VaR: Sample size 1000.

y	Empirical	VaR$_{inf}$	VaR$_{sup}$	True density	SME
0.950	3.2360580	3.1938743	3.2458468	3.2006401	3.2486497
0.960	3.2473619	3.2283227	3.2856532	3.2326465	3.2886577
0.970	3.3106142	3.2473619	3.3283750	3.2726545	3.3366673
0.975	3.3220161	3.2770023	3.3754777	3.3046609	3.3686737
0.990	3.4239602	3.3837184	3.4385691	3.4246849	3.5287057
0.995	3.4717304	3.4239602	3.5427976	3.5127025	3.6407281
0.999	3.6898073	3.4717304	3.6898073	3.6967393	3.8887778

Table 12.8: TVaR: Sample size 1000.

y	Empirical	TVaR$_{inf}$	TVaR$_{sup}$	True density	SME
0.950	3.3547538	3.2968652	3.4185504	3.3379840	3.4191404
0.960	3.3821343	3.3181586	3.4559391	3.3684504	3.4572513
0.970	3.4191905	3.3443793	3.4966896	3.4067017	3.5056193
0.975	3.4394590	3.3600881	3.5232593	3.4303656	3.5358641
0.990	3.5373198	3.4138775	3.6834803	3.5437671	3.6842614
0.995	3.6208459	3.4291698	3.8357996	3.6246247	3.7935868
0.999	3.8343980	3.4471884	3.9789886	3.8002637	4.0422850

We only add that similar tables corresponding to the examples in Section 8.4.2, in which the data was contaminated with a small measurement error, yield very similar results, which we refrain from presenting to the reader.

12.2.2 VAR and TVaR of aggregate losses

In this section we show the results of the computation of the VaR and TVaR of the densities of losses that comprise several levels of aggregation.

Independent and partially coupled risks
The numerical example of density determination for this case was treated in Section 7.5.2. Let us begin with the first case depicted in Figure 7.5. To have a few numbers as references for comparisons, we first compute the empirical VaR and TVaR for each case using a large sample of size 9000 at confidence levels equal to 95 % and 99 %. The result is presented in Table 12.9.

Besides each empirical value, in Table 12.9 we show as well the confidence intervals obtained by resampling (or bootstrapping) with a large subsample.

Table 12.9: VaR, TVaR and confidence intervals for the simulated S of size 9000.

S	VaR				TVaR			
	0.95		0.99		0.95		0.99	
	Emp.	Conf. int.	Emp.	Conf. int.	Emp.	Conf. int.	Emp.	Conf. int.
	7.877	7.849–7.910	8.601	8.560–8.652	8.301	8.239–8.363	8.888	8.799–8.979

For the comparison to be fair, in this case we computed the integral (12.3) using as the upper limit of integration the maximum value of the sample. Since the maximum entropy method provides us with an analytic expression for the density, we can use a much larger interval to obtain a better estimate of the VaR and the TVaR.

In Table 12.10 we show the result of computing the VaR and the TVaR for the densities corresponding to Figure 7.5. If need be, the reader should go back to Chapter 7 to review how the different densities plotted in that figure were obtained.

Table 12.10: Comparison of VaR and TVaR for the SME, and convolution approaches for several cases.

Copula	VaR				TVaR			
	0.95		0.99		0.95		0.99	
	SME	Conv.	SME	Conv.	SME	Conv.	SME	Conv.
Independent	7.793	7.828	8.698	8.603	8.527	8.332	9.190	8.922
Gaussian, $\rho = 0.5$	7.560	7.672	8.045	7.983	7.868	7.850	8.272	8.292
Gaussian, $\rho = 0.8$	7.682	7.672	8.136	8.138	7.9568	7.978	8.379	8.435
t-Student, $\rho = 0.7$, $\nu = 10$	7.672	7.707	8.138	7.983	7.936	7.914	8.348	8.317

Fully coupled risks

This case corresponds to the second example treated in Section 7.5.2. The densities obtained in this case are plotted in Figure 7.6. In Table 12.11 we present the VaR at various confidence levels (indicated in the first column), obtained from the maxentropic densities as with the densities obtained by the copula based convolution approach. In the last two columns of Table 12.11 we list the endpoints of the confidence interval for the different values of γ, computed at the 95 % confidence level by bootstrapping from the empirical data.

The description of Table 12.12 is similar to that of Table 12.11, so there is no need to add much to this. We end this section by mentioning the results of computing the VaR and the TVAR at various levels of confidence using (12.3) with $10 \max(S)$ as indicated there.

12.3 Computation of risk premia

Actually, the thrust of this section is not the computation of the risk premia, but to examine the dependence of the risk premia on the sample size. In our case this is easy

Table 12.11: Comparison of VaR for the SME, and convolution approaches.

y	Approaches			Errors		Confidence interval 95 %	
	SME	Conv.	Empirical	SME err.	Conv. err.	VaR$_{inf}$	VaR$_{sup}$
0.900	7.431	7.569	7.491	0.060	0.078	7.442	7.522
0.950	7.707	7.707	7.704	0.003	0.003	7.641	7.757
0.990	8.259	8.259	8.231	0.028	0.028	8.102	8.603
0.995	8.397	8.397	8.672	0.275	0.275	8.231	8.804
0.999	8.810	8.672	8.999	0.189	0.327	8.689	9.065

Table 12.12: Comparison of TVaR for the SME, and convolution approaches.

y	Approaches			Errors		Confidence interval 95 %	
	SME	Conv.	Empirical	SME err.	Conv. err.	TVaR$_{inf}$	TVaR$_{sup}$
0.900	7.817	7.834	7.836	0.019	0.002	7.718	7.963
0.950	8.032	8.030	8.091	0.059	0.061	7.922	8.262
0.990	8.498	8.408	8.769	0.271	0.361	8.344	8.974
0.995	8.687	8.499	8.875	0.188	0.371	8.473	9.147
0.999	9.092	8.896	9.160	0.068	0.264	8.650	9.254

because the maxentropic procedure produces a density, which is consistent with the data even when the data is scarce. This problem was analyzed in the previous section. Here we just examine the influence of the variability of the reconstructed density on the expected values calculated using those densities.

The few definitions that we consider were taken from [59]. We direct the reader to chapter 5 in that volume in which premium principles are explained. In what follows, S will denote the total severity that the insurance company has to cover.

Consider to begin with the mean value principle: Let $U(s)$ be a concave function defined on $(0, \infty)$ and define

$$\pi(S) = U^{-1}(E[U(S)]). \tag{12.4}$$

Here U^{-1} stands for the compositional inverse of U. In utility theory, that quantity can be thought of as the certain equivalent of the risk S. Note that according to the material in Section 12.1,

$$E[U(S)] = U(0)P(S = 0) + (1 - P(S = 0)) \int f_S(s)U(s)ds.$$

When $U(0) = 0$ and only when $P(S = 0)$ can be neglected relative to 1, we obtain

$$\pi(S) = U^{-1}\left(\int f_S(s)U(s)ds\right). \tag{12.5}$$

As we want to illustrate the variability of the $\pi(S)$ due to the variability of the f_S, let us use two of the examples contained in the list (12.5) and suppose that our utility function is either of the two following cases:

$$U(x) = 1 - e^{-as} \quad \text{or} \quad U(x)\sqrt{x}.$$

Then, according to (12.5) the risk premia in each case are, respectively, given by

$$\pi(S) = -\frac{1}{a}\ln\left(\int f_S(s)e^{-as}ds\right) \quad \text{or} \quad \pi(S) = \left(\int_0^\infty \sqrt{(s)}f_S(x)\right)^2 dx.$$

Besides that we may also consider the stop loss with a cap given by

$$\pi(S) = E[\max(K,(S-d)^+)] = \int_d^{K+d}(s-d)f_S(s)ds + K\int_{K+d}^\infty f_S(s)ds.$$

Now we use the densities plotted in Figure 10.2, in which the samples sizes were $M = 10, 20, 50, 100, 500, 1000$, to compute $\pi_M(S) = E_{f_M}[\max(K,(S-d)^+)]$ for each f_M, and then we plot the results in a bar diagram.

In Figure 12.1 we present the results computing the risk premiums as the certain equivalent of the two possible utility functions mentioned above, as well as that premium with stop loss and cap.

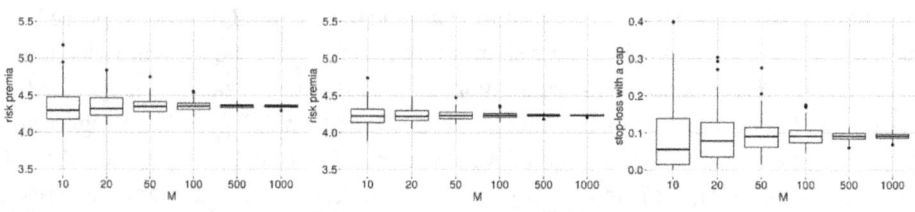

(a) Utility function $1 - e^{-as}$ (b) Utility function $\sqrt{(s)}$ (c) Stop loss with a cap

Figure 12.1: Premium variability for different utilities.

As indicated in the figure, the left panel describes the variability of the premium for the exponential utility function, the center panel that of the square root function, and the right panel the stop loss premium. As expected, the boxes shrink when the sample size increases. The data used for the figures is as follows: For the exponential utility function we used $a = 1$. We considered $d = E[S] + \sigma_S$, and we set $K = \text{VaR}_{95}(S)$. In each case, that is for every sample size, we use the corresponding density f_M to compute these values.

13 A solution to the capital allocation problem

13.1 Introduction and preliminaries

In the previous chapters, we saw how to extract information from loss data in the form of loss densities. The losses may be thought about in several possible ways. They may be operational risk losses due to different risk factors at a financial institution, or losses due to different types of coverage sold by an insurance or reinsurance company, or losses prone to happen in the activities of some enterprize.

The corporation may use the loss aggregation methodology to determine a risk capital to cover the potential loss in each line of work and for the aggregate risk. Then the problem that comes up is as follows: How much risk capital to allocate to each line of activity when the total money for hedging the aggregate risk is fixed?

The capital allocation problem is stated as follows: Consider a collection of risks $\{X_i, i = 1, \ldots, N\}$ modeled by continuous positive random variables and adding up to a total risk $\sum X_i$. Given a total risk capital K, we want to determine the capitals $\{K_i, i = 1, \ldots, N\}$ to hedge each risk in such a way that $\sum K_i = K$.

Throughout we assume once again that $\boldsymbol{X}$ is a $[0, \infty)^N$-valued random variable with some underlying distribution density $F_{\boldsymbol{X}}(\boldsymbol{x})$. The joint density will play no role in the numerical approach that we develop to solve the allocation problem. We do not make use of these (generally) unknown distributions because the maxentropic method relies either on expert opinion or on bounds for the standalone risks determined as explained in the previous chapters.

There exists a theoretical proposal to characterize a capital allocation procedure that goes as follows. First, choose a risk measure ρ defined on the class of risks under consideration. A capital allocation rule subordinated to the risk measure ρ is specified as follows. Let $\{X_i, i = 1, \ldots, N\}$ be the individual risks, and let $X = \sum_{i=1}^{N} X_i$ be the aggregate risk. An allocation rule assigning capital $\Lambda(X_i, X)$ to risk X_i has to satisfy $\sum_{i=1}^{N} \Lambda(X_i, X) = K$ (the total available capital), and for any possible risk Y, it allocates its standalone price of risk, that is, $\Lambda(Y, Y) = \rho(Y)$.

Which risk measure and which capital allocation rule to apply is usually left up to the risk manager. The answer depends on the business sector in which the risk allocation process arises. For example, the banking sector uses VaR or TVaR as required by regulatory agencies. A manufacturing corporation might use them to estimate costs or losses. But there is neither a priori way to contrast different risk measure choices nor the result of capital allocation rules. This adds value to the numerical procedure that we present below, which does not make use either of a risk measure or of a capital allocation rule.

Problem 1: How to assign risk capital within specified bounds?
Suppose that the risk management department has no preferred risk measure and asked several experts to provide lower and upper bounds for the capital to be assigned to each

risk. The experts come up with a range $[L_i, U_i]$ within for the capital K_i to be assigned to the ith risk.

Experts may suggest to use $U_i = \text{VaR}_{0.99}(X_i)$ or perhaps $U_i = \text{TVaR}_{0.975}(X_i)$ as an upper bound, and for the lower bound, they may suggest using the actuarial risk price $E[X_i]$ of each risk.

Thus, at this stage, all that the risk analyst knows or is provided with is a range $[L_i, U_i]$ in which the risk capital K_i is to be allocated. Since $U_i - L_i$ can be interpreted as the unexpected loss incurred by the ith risk, and the risk manager being optimistic, she assigns $K < \sum U_i$ to cover all the potential losses. With this, the risk capital allocation problem to be solved can be stated as follows:

$$\text{Solve} \quad \sum_{i=1}^{N} K_i = K \quad \text{subject to} \quad L_i < K_i < U_i \quad \text{for } i = 1, \ldots, N. \tag{13.1}$$

This is an ill-posed linear algebraic problem consisting of one equation with many unknowns with a convex (box) constraint upon the solutions. In Chapter 8, we showed how to solve this problem using the method of maximum entropy in the mean. We will see how to adapt it in the numerical example below.

Problem 2: Determination of the risk measure from the market prices of risk

Since the upper bounds $\{U_i, i = 1, \ldots, K\}$ for the individual risks can be taken to be the standalone risks prices for the risks, we are led to another interesting problem. If we suppose that the risk measure that yields the U_i as risk prices is a spectral risk measure, then we can apply a maximum entropy procedure to determine it. The potential use of this possibility is that the risk measure so obtained can be used to value further risks. We settle on the spectral risk measures, first, because they are coherent, and second, because they lead to a tractable numerical problem. We briefly sketch the basics about spectral risk measures in the Appendix to this chapter. The inverse problem may be stated as

$$\text{Determine} \quad \phi(u) \quad \text{such that} \quad \int_0^1 \phi(u) \text{VaR}_u(X_i) du = U_i, \quad i = 1, \ldots, N. \tag{13.2}$$

To lead to a coherent risk measures, the function ϕ has to be positive, increasing, and subject to the integral constraint $\int_0^1 \phi(u) du = 1$. Clearly, (13.2) is an ill-posed linear problem subject to convex constraints. After discretization, this infinite-dimensional problems becomes finite-dimensional, but it is still ill-posed.

The maxentropic methodology that we use to solve this second problem was originally used by Gzyl and Mayoral [48]. For us, the interest lies in the intertwining of the two problems. Once Problem 2 is solved for a class of risks, the ρ so determined can be used to determine the range for the capital allocation of any other collection of risks, and the procedure developed to solve Problem 1 can be brought in to determine the capital allocation problem for the new collection of risks.

13.2 Numerical results

The numerical examples considered below were treated by Gzyl and Mayoral [49].

First, we work out an example of a constrained capital allocation problem. After that, we consider three examples of determination of a distortion function from risk prices. Along the way, we examine how the two procedures tie up.

The examples are chosen to appear realistic, and they do not correspond to any real bank or corporation or anything. The parameters were chosen arbitrarily. The bounds on the risks reflect a possible choice by a risk manager. The different cases considered in the second (sub)section are chosen to illustrate the potential of the maximum entropy in the mean to reproduce prices of statistically diverse risks by means of one single distorted risk measure and its possible use for risk pricing for capital allocation.

13.2.1 The capital allocation

To obtain upper and lower bounds for the capital allocation problem, suppose the standalone risks are known to be log-normal random variables with parameters (μ, σ), that is, $X = \exp(\mu+\sigma\zeta)$ with $\zeta \sim N(0,1)$. We suppose that the risk manager considers $L = E[X]$ as a lower bound for all risks but has several possible choices for the upper bound: $U = \text{VaR}_{0.95}(X)$, $U = \text{VaR}_{0.99}(X)$, $U = \text{TVaR}_{0.95}$, or $U = \text{TVaR}_{0.975}$.

For log-normal random variables, the values of U are obtained from their analytical expressions, namely

$$\text{either} \quad U = \text{VaR}_\alpha(X) = e^{\mu+z_\alpha\sigma} \quad \text{or} \quad \text{TVaR}_\alpha(X) = \left[\frac{\Phi(\sigma - z_\alpha)}{1-\alpha}\right]e^{\mu+\sigma^2/2}.$$

Here z_α denotes the 100α-quantile of the $N(0,1)$ random variable, and Φ denotes its cumulative distribution function.

The input information for the numerical work is summarized in Table 13.1. There we display the parameters of the distribution plus the lower and upper bounds of the different risks. To study the effect of the coefficient of variation (CV), we organize the data as follows. In panel A, we fixed $\sigma = 0.1$ (which in this case means fixing the coefficient of variation) and let μ vary as indicated, and in panel B, we fixed μ, and let σ (or CV) vary. This describes what affects the bounds of the risks. In both panels the columns labeled by Σ contain the sum of the lower or upper bounds of the risk capitals. Even though the computations were carried out at high precision, we only report figures with two decimal places for the table to fit on the page. The row labeled L contains the mean of each risk, and the rows labeled VaR and TVaR list two possible upper bounds for each risk at the specified confidence levels. The knowledge of the sum of the lower or upper bounds allows management to choose the total capital to be allocated as a value between these values. To solve numerically the capital allocation problem, for each risk level, we

Table 13.1: Input data for the capital allocation problem.

	PANEL A					PANEL B				
	X_1	X_2	X_3	X_4	Σ	X_1	X_2	X_3	X_4	Σ
σ	0.10	0.10	0.10	0.10		0.30	0.50	0.70	1.00	
μ	1.00	1.50	2.00	2.50		1.00	1.00	1.00	1.00	
L	2.73	4.50	7.43	12.24	26.91	2.84	3.08	3.47	4.48	11.03
$VaR_{0.95}$	3.20	5.28	8.71	14.36	31.56	4.45	6.17	8.86	14.08	33.32
$VaR_{0.99}$	3.43	5.66	9.32	15.37	33.78	5.46	8.70	13.85	27.84	55.85
$TVaR_{0.95}$	3.50	5.77	9.51	15.57	34.44	5.64	8.81	13.30	23.26	51.01
$TVaR_{0.975}$	3.59	5.93	9.77	16.11	35.40	6.12	10.07	16.02	30.21	62.43
CV	0.10	0.10	0.10	0.10		0.31	0.53	0.80	1.31	

considered two possible values of the total capital to be allocated, ranging between the sums of the lower and upper bounds. The results obtained are shown in Table 13.2.

Table 13.2: Allocated risk capitals.

	PANEL A					PANEL B				
	K	K_1	K_2	K_3	K_4	K	K_1	K_2	K_3	K_4
$VaR_{0.95}$	29.50	2.98	4.92	8.13	13.47	31.00	3.88	5.43	7.90	13.79
$VaR_{0.95}$	31.00	3.05	5.11	8.55	14.29	32.00	3.98	5.70	8.30	14.03
$VaR_{0.99}$	29.50	3.05	5.00	8.17	13.27	31.00	4.11	5.71	8.05	13.12
$VaR_{0.99}$	31.00	3.10	5.14	8.53	14.23	32.00	4.12	5.76	8.21	13.91
$TVaR_{0.95}$	32.00	3.16	5.26	8.79	14.79	49.00	4.84	8.03	12.91	23.22
$TVaR_{0.95}$	34.00	3.32	5.61	9.42	15.66	50.00	4.07	8.46	13.21	23.26
$TVaR_{0.975}$	32.00	3.32	5.29	8.80	14.72	49.00	4.63	7.25	11.86	25.26
$TVaR_{0.975}$	34.00	3.28	5.53	9.36	15.83	50.00	4.65	7.33	12.09	25.93

The results in Table 13.2 are organized as follows: In each panel, in each case the lower bound for each risk is the value of L listed in Table 13.1. The first four rows of Table 13.2 correspond to a total risk capital larger than L but smaller than U obtained by summing the individual values of the VaR, and the last four rows correspond to the total risk larger than L but smaller than the sum of the individual values of TVaR. The two panels display the effect of a constant coefficient of variation or a variable coefficient of variation. They also show the effect of varying total risk capital, listed in the column labeled K in each panel of Table 13.2.

When the allocated capital lies exactly at the midpoint between the minimum and maximum, the resulting allocation is also at the midpoint between L_i and U_i as must be clear after a glance at (10.5). Also, the method becomes unstable when the total capital to be allocated is chosen to coincide with any of its extreme values. When the constraint is at the boundary of the allowed values, the minimization of (13.3) becomes numerically unstable.

13.2.2 Problem 2: Determining the distortion function from given risk prices

Recall that the purpose of this section is to guess how a corporation may use the market prices of risk to determine a risk measure to price its risks according to the market.

We suppose that assets are distributed according to either a Lognormal, a Gamma or a Pareto density. The first two are quite common in the insurance industry, but nevertheless, consider Kiche et al. [60]. Consider for example the recent work by Park and Kim [79], Jakata and Chikobvu [56], in which the Pareto (standard or generalized), is used in risk management and in applications of Extreme Value Theory for estimating large losses and tail risk. To finish, consider an application of distorted risk measures to systemic risk developed in Dhaene et al. [28].

As mentioned at the beginning of Section 13.1, the upper bound of the range for the capital allocation problem may be considered to be its standalone risk price. All that the analyst knows is the statistical distribution of the different risks and that (*as the risk manager assumes*) the expert uses some distortion function to price the risks, and the problem is to determine that distortion function. There are two important reasons behind this essential assumption. On the theoretical side, distorted risk measures are a large and flexible class of coherent measures, and from the practical side, they lead to a problem that can be solved by the MEM methodology.

Once the distortion function is obtained, we can use it to price other risks. Not only that, we can use this risk price to determine ranges for the capital to be assigned to a new collection of risks using the methodologies proposed in Section 13.3 and the previous example.

For the computations that we describe below, we consider the risks distributed according to:
(a) generalized Pareto with density $f(x; k\sigma, \theta) = \frac{1}{\sigma}(1 + k\frac{(x-\theta)}{\sigma})^{-(1+1/k)}$,
(b) gamma with density $f(x; a, b) = \frac{1}{b^a \Gamma(a)} x^{a-1} e^{-x/b}$, and
(c) log-normal with density $f(x; \mu, \sigma) = \frac{1}{x\sigma\sqrt{2\pi}} e^{-(\ln x - \mu)^2 / 2\sigma^2}$.

Before describing the numerical examples, we mention at this point that to assess the effect of the size of the domain, in all the numerical experiments, we used partitions of $[0, 1]$ of sizes $n = 20, 50, 100, 200$. All solutions look alike, and we report the cases $n = 50$ and $n = 100$ only. Recall that for this, we apply MEM to solve two ill-posed problems consisting of solving a set of 4 or 6 equations to determine 50 or 100 unknowns. In all the plots displayed below, the dashed curve is that of the true distortion function, and the boxes lie on the reconstructed (estimated) distortion function. Of course, in actual practice the original function is not known. This is done here to illustrate the performance of MEM.

Different risk distributions

In the first example, we consider six risks characterized by the distributions specified in Table 13.3. As already said, the analyst wants to determine the distortion function used by the expert to value the risks. The relationship between the distortion functions used as inputs for MEM was explained in Section 13.3.3 of Appendix, and a few remarks about distortion functions are presented in Section 13.3.2 of Appendix. We suppose the expert used a proportional hazard distortion function $g(u) = u^{1/\gamma}$ with $\gamma = 1.5$ to compute the risk prices listed in the first row of Table 13.4. In the second row, we list the prices computed with the distortion function obtained by applying MEM. As seen in Table 13.4, the agreement between the input data and predicted prices is good up to the third decimal place. In Figure 13.1, we display the original (dashed line) and the numerically reconstructed distortion function (dotted line) for two different mesh sizes.

Table 13.3: Risk densities and their parameters.

Distribution	k	σ	θ	CV
Pareto 1	0.25	1.00	1.00	1.52
Pareto 2	0.25	2.00	20	3.04
	–	a	b	CV
Gamma 1		1.50	2.50	0.82
Gamma 2		1.00	2.00	1.00
		μ	σ	CV
LogNormal 1		0.50	0.40	0.42
LogNormal 2		1.00	0.70	0.80

Table 13.4: Given and determined prices.

True Price	2.5953	5.2107	4.765	2.4171	2.0309	4.0483
Estimated Price	2.5929	5.2114	4.7647	2.4181	2.0314	4.0483

As mentioned above, the distortion function just computed can be used to compare the risk prices of a new set of risks, or perhaps to compare them with the risk prices that the hired expert provides us. The sanding assumption is that she/he always uses the same method to compute risk prices. So, let us consider the risks described in Table 13.5.

Given this data, we compute the risk prices using the exact (original) distortion function (that is, the one that the expert would have used) and the distortion function determined numerically using the previous data set. In the first row of Table 13.6, we list the prices that the expert would provide us, and in the second, we list the prices computed with the maxentropic distortion function computed previously. We emphasize that this is an indirect performance test since we know the true prices because we know the distortion function that the expert uses.

Keeping in mind that we are using 6 values of an integral to reconstruct at least 50 points of a density, the agreement is quite good.

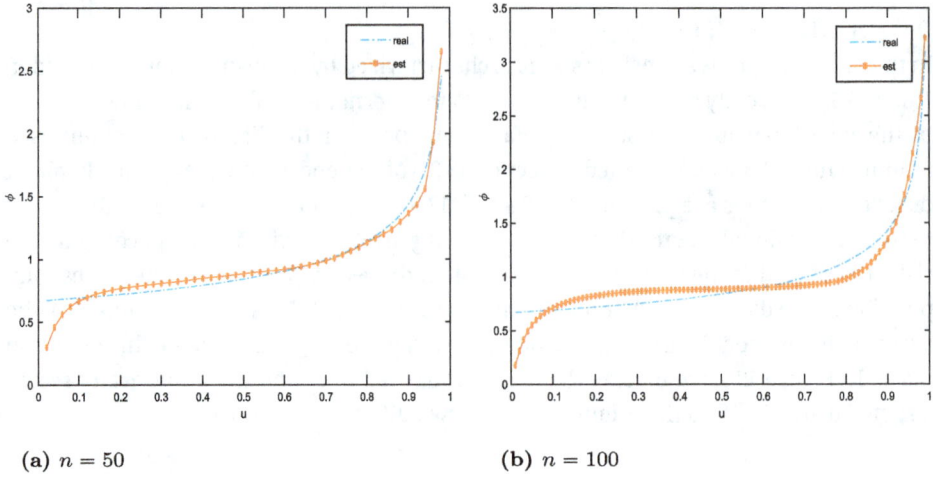

(a) $n = 50$ (b) $n = 100$

Figure 13.1: True and estimated distortion functions.

Table 13.5: Risk densities and their parameters.

Distribution	k	σ	θ	CV
Pareto 1	0.30	1.30	2.60	**1.93**
Pareto 2	0.30	2.50	5.00	**3.70**
	–	a	b	CV
Gamma 1		1.25	2.25	**0.89**
Gamma 2		1.750	2.75	**0.76**
		μ	σ	CV
LogNormal 1		0.70	0.60	0.66
LogNormal 2		1.00	0.70	**1.12**

Table 13.6: A robustness test: Given and determined prices.

	X_1	X_2	X_3	X_4	X_5	X_6	Σ
True Price	5.2027	9.2242	3.5862	5.6455	2.8779	5.7918	32.3283
Estimated Price	5.2234	9.2281	3.5667	5.6383	2.8749	5.7981	32.3295

To close the circle, we apply the technique proposed in the first example to solve the capital allocation problem for the new collection of risks. That is, as lower bounds for the capital allocation, we take their means, and as upper bounds, their risk prices (from Table 13.6) thought of as standalone risk prices. For values of the total risk capital between the sums of the lower and upper bounds, we compute the allocated capitals per risk. The results are displayed in Table 13.7

Table 13.7: Given and determined prices.

	X_1	X_2	X_3	X_4	X_5	X_6	Σ
Lower bound	4.4571	8.5714	2.8125	4.8125	2.4109	4.5042	27.5686
Upper bound	5.2234	9.2281	3.5667	5.6383	2.8749	5.7981	32.3295
Alloc. Cap.	4.9939	9.0144	3.3388	5.4023	2.7014	5.5493	31
Alloc. Cap.	5.0763	9.0798	3.4193	5.4936	2.738	5.6931	31.5
Alloc. Cap.	5.1674	9.1613	3.5096	5.5881	2.793	5.7806	32

All risks from the same family of distributions

To better examine the role of the statistical nature of the risks and that of the distortion function, we consider three sets of risks, with Pareto, gamma, and lognormal densities, and suppose that two experts price using two distortion functions, a proportional hazard and the Wang distortion function. The proportional hazard distortion function is as above, and the Wang distortion is $g(u) = \Phi(\Phi^{-1}(u) + \lambda)$ with $\lambda = 0.05$; see Wang [97] for more detail on this class of distortion functions. We will use these as data to determine the distortions and examine how they price the related risks.

To examine the effect of the statistical nature of the risks, we will suppose that all risks in the data set have the same coefficient of variation but that the new risks to price have all different coefficients of variation. In all cases, we use a partition of size $n = 100$ to discretize $[0, 1]$, and we will split the example into three cases according to the nature of the risk.

For typographical reasons, we organize the description in two cases. First, we consider the Pareto case in detail, and then we consider the gamma and the lognormal cases. The description of the tables and plots is the same as above, and the descriptions are shorter.

Case 1: Pareto risks

The Pareto risks have the heaviest tails. The first data set is described in Table 13.8. In

Table 13.8: Risk densities and their parameters.

Distribution	k	σ	θ	CV
Pareto 1	0.25	0.3	0.30	0.46
Pareto 2	0.25	0.5	2.50	0.76
Pareto 3	0.50	0.75	0.75	0.87
Pareto 4	0.20	21.00	1.00	1.16

the first and third rows of Table 13.9, we list the prices of the risks used as inputs for the application of MEM to determine the distortion functions. The distortion functions are shown in Figure 13.2.

Table 13.9: A consistency test: original versus reconstructed prices.

True Price (Pareto) (PH)	0.791	1.4789	2.1313	2.6408
Estimated Price (Pareto) (PH)	0.7916	1.7496	2.13	2.6413
True Price (Pareto) (Wang)	1.2555	2.0055	2.8704	3.7334
Estimated Price (Pareto) (Wang)	1.2551	2.0066	2.8684	3.7345

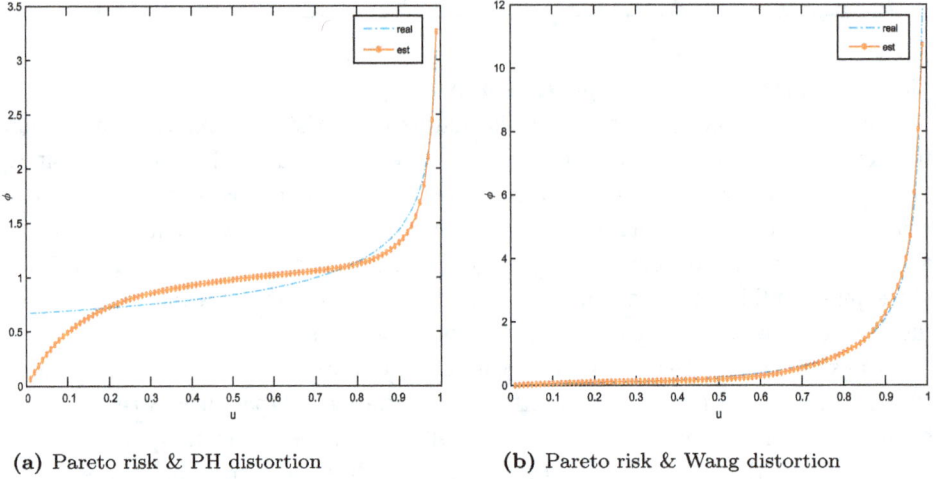

(a) Pareto risk & PH distortion   (b) Pareto risk & Wang distortion

Figure 13.2: True and estimated distortion functions.

This time the reconstructed distortion functions (shown by dotted lines) seem to cling better to the "true" distortion (shown by a dashed line). This may, perhaps, be due to the fact that the distributions of all the risks belong to the same family. The test of consistency this time yields the results displayed in Table 13.9.

In the even rows of Table 13.9, we list the prices determined using the reconstructed distortion function. The first consistency test passed reasonably well. To carry out the second, we again used the original distortion function to compute the "true" prices and then the reconstructed function to compute the "predicted" prices. In Table 13.10 we display the data about the new risks to be priced, and in the last table, we display two sets of prices.

Table 13.10: Risk densities and parameters of test case.

Distribution	k	σ	θ	CV
Pareto 1	0.15	1.00	1.00	0.91
Pareto 2	0.40	0.50	0.50	2.60
Pareto 3	0.20	2.00	1.50	1.16
Pareto 4	0.30	1.75	1.50	4.46

To finish the list of tables for this case, in Table 13.11 we display the predicted price of a new collection of risks along with the prices of the same risks calculated using the original (true) distortion function.

Table 13.11: A robustness test: original versus predicted prices.

True Price (Pareto) (PH)	2.4837	1.5951	2.7469	4.5039
Estimated Price (Pareto) (PH)	2.4908	1.5977	2.7547	4.4930
Estimated Price (Pareto) (Wang)	3.6440	2.6218	3.2840	6.7449
Estimated Price (Pareto) (Wang)	3.6410	2.6070	3.2922	6.7125

Cases 2 and 3: Gamma and lognormal risks

The data for these examples are presented in Table 13.12.

Table 13.12: Risk densities and their parameters.

Distribution	σ	θ	CV
Gamma 1	1.25	1.50	0.46
Gamma 2	1.50	1.75	0.76
Gamma 3	2.75	2.00	0.87
Gamma 4	2.25	2.50	1.16
	μ	σ	CV
Lognormal 1	0.50	0.44	0.467
Lognormal 2	1.00	0.68	0.76
Lognormal 3	1.25	0.75	0.87
Lognormal 4	1.50	0.92	1.16

This yields the reconstructed distortion functions displayed in Figure 13.3. The dashed line corresponds to the original distortion, and the dotted line to the reconstructed one.

The first consistency test yields Table 13.13. Again, odd rows show the prices used as data, and the even rows show the "predicted" prices, and again, this measures the quality of the reconstruction.

To double check on the applicability of the reconstructed distortion function, we consider the data with variable coefficient of variation given in Table 13.10.

The second consistency test applied to this new data set yields Table 13.15. The description of the entries is as above, the odd rows contain the original data, and the even rows contain the predicted data.

Before closing this section, we mention again that the test of the robustness of the reconstructed distortion function, in which the prices of new risks are computed with the reconstructed density, cannot be performed in real life because the true prices are actually unknown.

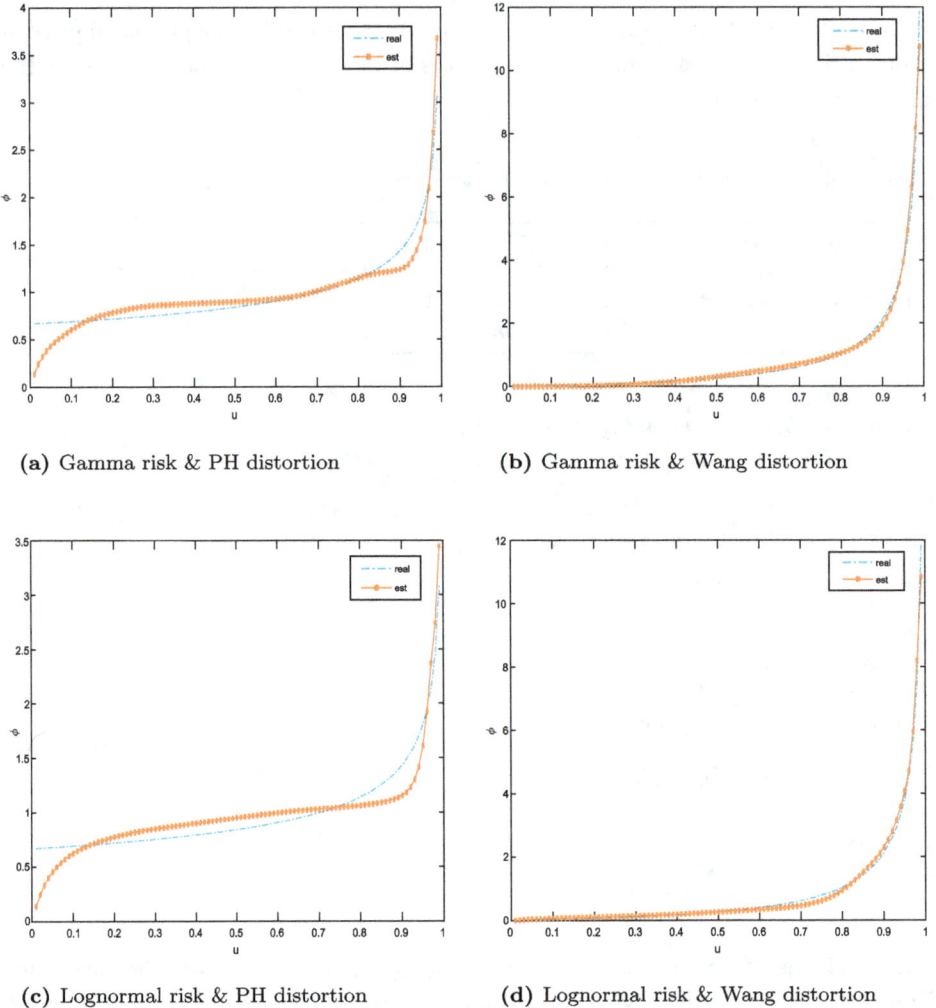

(a) Gamma risk & PH distortion (b) Gamma risk & Wang distortion

(c) Lognormal risk & PH distortion (d) Lognormal risk & Wang distortion

Figure 13.3: True and estimated distortion functions.

13.3 Appendices

13.3.1 Application of MEM to determine the capital allocation

Problem (13.1), consists of determining numbers K_i such that

$$\sum_{i=1}^{N} K_i = K \quad \text{under the constraint} \quad L_i \leq K_i \leq U_i \quad \text{for } i = 1, \ldots, N.$$

Comparing with the notations of Section 13.2, we see that here $n = N$ and $d = 1$. In this case, $\boldsymbol{A} = \boldsymbol{u}^t$ is an N-dimensional row vector with all components equal to 1, and to

Table 13.13: A consistency test: original versus reconstructed prices.

True Price (Gamma) (PH)	10.9353	4.4893	4.4139	3.0414
Estimated Price (Gamma) (PH)	10.9364	4.4881	4.4123	3.0430
True Price (Gamma) (Wang)	12.3043	6.8233	6.1334	4.5165
Estimated Price (Gamma) (Wang)	12.3043	6.8233	6.1334	4.5165
True Price (Lognormal) (PH)	2.0670	3.9276	5.1457	10.3495
Estimated Price (Lognormal) (PH)	2.0689	3.9271	5.1453	10.3496
True Price (Lognormal) (Wang)	2.4675	6.1616	7.9963	14.9900
Estimated Price (Lognormal) (Wang)	2.4662	6.1619	7.9968	14.9898

Table 13.14: Risk densities and parameters of test case.

Distribution	σ	θ	CV
Gamma 1	1.10	1.75	1.0
Gamma 2	1.20	2.00	0.92
Gamma 3	1.50	2.25	0.82
Gamma 4	2.00	2.50	0.71
	μ	σ	CV
Lognormal 1	0.70	0.30	0.31
Lognormal 2	1.20	0.70	0.80
Lognormal 3	1.40	0.90	1.12
Lognormal 4	1.60	1.10	1.53

Table 13.15: A robustness test: original versus predicted prices.

True Price (Gamma) (PH)	2.3637	2.9110	4.0784	5.901
Estimated Price (Gamma) (PH)	2.3505	2.9138	4.0853	5.8917
True Price (Gamma) (Wang)	3.1863	4.8174	6.2996	8.6080
Estimated Price (Gamma) (Wang)	3.1835	4.8043	6.2970	8.6115
True Price (Lognormal) (PH)	2.260	5.3935	7.5330	11.4069
Estimated Price (Lognormal) (PH)	2.2670	5.3595	7.4887	11.3954
True Price (Lognormal) (Wang)	2.4307	7.5638	14.1933	23.2203
Estimated Price (Lognormal) (Wang)	2.4298	7.5450	14.1145	22.9311

finish, the right-hand side of (8.1) becomes just a real number $y = K$. The constraint set for this situation is $\mathcal{C} = \prod_{i=1}^{N}[L_i, U_i]$. As a measure Q on $\mathcal{C}$ such that the convex hull of its support is $\mathcal{C}$, we consider

$$dQ(\xi) = \prod_{i=1}^{N}(\epsilon_{L_i}(d\xi_i) + \epsilon_{U_i}(d\xi_i)),$$

where $\epsilon_a(d\xi)$ denotes the measure that puts a unit mass at point a. The intuition behind the choice is that any point with an interval $[L, U]$ is a convex combination of its end points.

With that choice of Q, the computation is simple:

$$Z(\lambda) = \int_C e^{-\lambda\langle u,\xi\rangle} dQ(\xi) = \prod_{i=1}^{N}(e^{-\lambda L_i} + e^{-\lambda U_i}).$$

To complete the procedure, we must minimize the dual entropy, which in this case is

$$\Sigma(\lambda, K) = \sum_{i=1}^{N} \ln(e^{-\lambda L_i} + e^{-\lambda U_i}) + \lambda K. \tag{13.3}$$

Once the minimizer λ^* is determined, an application of (10.5) to compute of the solution to (13.1) yields

$$K_i^* = L_i \frac{e^{-\lambda^* L_i}}{e^{-\lambda^* L_i} + e^{-\lambda^* U_i}} + U_i \frac{e^{-\lambda^* U_i}}{e^{-\lambda^* L_i} + e^{-\lambda^* U_i}}. \tag{13.4}$$

Observe that:
(i) The weights in the convex combination depend only on the available information.
(ii) When all risks fall into the same range, say $[L, U]$, then all allocated capitals are the same and equal to K/N, because λ^* is determined so that this constraint is satisfied.
(iii) If the business units have very small unexpected risks, that is, if $L_i \approx U_i$, then the allocated capitals are $K_i^* \approx L_i$.
(iv) Observe also that (13.4) allows the analysis of the sample dependence of the solution on the estimate of the lower and upper limits on the range of the capital.

To analyze the relationship between λ and the total capital K to be allocated, notice that the first-order condition determining λ^* is

$$\frac{d\Sigma(\lambda, K)}{d\lambda} = 0 = -\sum_{i=1}^{N} L_i \frac{e^{-\lambda^* L_i}}{e^{-\lambda^* L_i} + e^{-\lambda^* U_i}} + U_i \frac{e^{-\lambda^* U_i}}{e^{-\lambda^* L_i} + e^{-\lambda^* U_i}} + K.$$

If we now differentiate the last identity with respect to K and isolate $d\lambda^*/dK$, then we obtain

$$\frac{d\lambda^*}{dK} = -\frac{1}{C(\lambda^*)},$$

where $C(\lambda^*) > 0$ is given by

$$C(\lambda^*) = \sum_{i=1}^{N} \left(L_i^2 \frac{e^{-\lambda^* L_i}}{e^{-\lambda^* L_i} + e^{-\lambda^* U_i}} + U_i^2 \frac{e^{-\lambda^* U_i}}{e^{-\lambda^* L_i}} \right) - \left(\sum_{i=1}^{N} L_i \frac{e^{-\lambda^* L_i}}{e^{-\lambda^* L_i} + e^{-\lambda^* U_i}} + U_i \frac{e^{-\lambda^* U_i}}{e^{-\lambda^* L_i}} \right)^2.$$

This happens to be a variance, and therefore it is positive. To conclude, we have the following:

Proposition 13.1. *With the notations introduced above, a glance at (13.4) shows that:*
(1) As $\lambda^* \to -\infty$, $K_i^* \to U_i$, and therefore $K \to \sum_{i=1}^N U_i$.
(2) As $\lambda^* \to \infty$ $K_i^* \to L_i$, and therefore $K \to \sum_{i=1}^N L_i$.

13.3.2 Distorted risk measure

A distortion function is defined as a positive increasing differentiable function $\phi : [0,1] \to [0,\infty)$ such that $\int_0^1 \phi(u)du = 1$. Define its associated (distorted) risk measure by

$$\rho(X) = \int_0^1 \phi(u) q_X(u) du, \tag{13.5}$$

where $q_X(u)$ is the quantile function of X. Since we suppose that our risks are modeled by continuous random variables with strictly positive density, we may also write $q_X(u) = F_X^{-1}(u)$. With the function $\phi(u)$, we can associate a distortion function $g : [0,1] \to [0,1]$ by means of $\phi(u) = g'(1-u)$ plus the conditions $g(0) = 0$ and $g(1) = 1$. In terms of g the risk price given by (13.5) can be now written in any of the two equivalent ways:

$$\rho(X) = \int_0^\infty x\, dg(F(x)) = \int_0^1 \phi(u) q_X(u) du. \tag{13.6}$$

To use a more financially suggestive notation, we will write either (13.5) or (13.6) as

$$\rho(X) = \int_0^\infty x\, dg(F(x)) = \int_0^1 \phi(u) \operatorname{VaR}_u(X) du, \tag{13.7}$$

where $\operatorname{VaR}_u(X) = q_X(u)$. Starting from $\operatorname{TVaR}_u(X) = E[X \mid X > \operatorname{Var}_u(X)]$ and integrating by parts, we obtain

$$\operatorname{TVaR}_u(X) = \frac{1}{1-u} \int_u^1 \operatorname{VaR}_s(X) ds.$$

For a deeper analysis of these quantities as risk measures, the reader may consider Fölmer and Schied [38].

13.3.3 Application of MEM to determine the distortion measure

Here we explain how to use the results in Section 8.1 to solve Problem (13.2), to determine distortion function from a collection of risk prices. The distortion function thus obtained can then be used to set an upper bound on risk prices for future capital allocation problems.

To proceed, the first step consists of discretizing the problem. For this we will closely follow Gzyl and Mayoral [48]. To simplify notations in (13.2), we use $q_i(u) = \text{VaR}_{X_i}(u)$ and write it as

$$\rho_\phi(X_i) = \int_0^1 q_i(u)\phi(u)du = U_i, \quad i = 1,\ldots,N+1,$$

where to accommodate the condition $\int_0^1 \phi(u)du = 1$, we add X_{N+1} such that $q_{N+1}(u) = 1$ and $U_{N+1} = 1$. Thus in this case, we will consider $d = N+1$ and $\mathbf{y}^t = (U_1,\ldots,U_N,1)$. Also, we have a methodological constraint imposed upon ϕ.

To proceed to the discretization stage, we consider the partition of $[0,1]$ at points $u_j = j/n$. The choice of n depends on the known variability of the $q_i(u)$ in $[0,1]$. Let us define the $(N+1) \times n$ matrix $\mathbf{B}$ by setting $B_{i,j} = q_i(u_j)/n$ for $i = 1,\ldots,N+1$ and $j = 1,\ldots,n$. Set $\phi(a_j) = \phi_j$, where $a_j = \frac{1}{2}(u_j + u_{j-1})$ and $u_0 = 0$. With all this, the discretized version of problem (13.2) can be restated as follows: Solve

$$\mathbf{B}\phi = \mathbf{y}, \quad \phi \in \mathcal{K}, \tag{13.8}$$

where the constraint set $\mathcal{K} \subset \mathbb{R}^n$ is a convex set defined in this case as

$$\mathcal{K} = \{(\phi_1,\ldots,\phi_n) \mid 0 \leq \phi_1 < \cdots < \phi_j < \phi_{j+1} < \cdots < \phi_n\}.$$

To simplify the description of the constraints, we set $\phi_1 = x_1, \phi_2 = x_1 + x_2, \ldots,$ and $\phi_n = x_n + \cdots + x_1$, or $\phi = \mathbf{T}x$, where $\mathbf{T}$ is the obvious lower diagonal matrix describing the change of coordinates. Setting $\mathbf{A} = \mathbf{BT}$, we can restate our discretized problem as

$$\mathbf{A}x = \mathbf{y}, \quad x \in \mathcal{C}, \tag{13.9}$$

where now the convex constraint set is $\mathcal{C} = [0,\infty)^n$, that is, the positive orthant in $\mathbb{R}^n$. Clearly, once the vector x is at hand, ϕ is easily recovered. To apply the maximum entropy procedure, we begin by specifying a reference measure Q on $\mathcal{C}$. In the notation of Section 13.2, we choose

$$dQ(\xi) = \prod_{j=1}^n \left(\sum_{n=1}^\infty \frac{1}{n!} \epsilon_n(d\xi_j) \right),$$

that is, of unnormalized Poisson measures of parameter 1 on $\mathcal{C}$. This choice makes the computation of $Z(\lambda)$ very simple. Clearly, the convex hull of Q is $\mathcal{C}$. Note that in this case, for $\lambda \in \mathbb{R}^d$, the function $Z(\lambda)$ is given as

$$Z(\lambda) = \int_{\mathcal{C}} e^{-\langle A^t\lambda, z\rangle} dQ(z) = \prod_{j=1}^{n}(\exp(1 - e^{-(A^t\lambda)_j})).$$

This time, we need to minimize

$$\Sigma(\lambda, y) = \sum_{j=1}^{n} \ln(\exp(e^{-(1-A^t\lambda)_j})) + \langle \lambda, y \rangle \tag{13.10}$$

with respect to $\lambda \in \mathbb{R}^n$. Once the vector λ^* is at hand, we have

$$x_j^* = e^{-(A^t\lambda^*)_j} \quad \text{for } j = 1, \ldots, n, \tag{13.11}$$

from which we can obtain $\phi(j)$ as indicated above. To close, we mention that to avoid overflow or underflow when dealing with exponentials, it is convenient to replace the problem $Ax = y$ by the problem $(\frac{1}{K}A)x = \frac{1}{K}y$ with some appropriate scaling factor K.

13.3.4 A numerical instability issue

To understand why the data cannot be near the boundary of the constraint region, we consider a simple problem of minimizing the one-dimensional version of $\Sigma(\lambda)$ of Section 13.3.

Theorem 13.1. *Consider the strictly convex function*

$$\Sigma(\lambda) = \ln(e^{-\lambda L} + e^{-\lambda U}) + \lambda y.$$

If $y \notin [U, L]$, then $\Sigma(\lambda) \to -\infty$ as $\lambda \to \pm\infty$, that is, $\Sigma(\lambda)$ is unbounded below and cannot be minimized.

The proof follows from the following remarks.

Lemma 13.1. *Consider the function $x(\lambda)$ defined on $\mathbb{R}$ as*

$$x(\lambda) = L\frac{e^{-\lambda U}}{e^{-\lambda L} + e^{-\lambda U}} + U\frac{e^{-\lambda U}}{e^{-\lambda L} + e^{-\lambda U}}.$$

Then $x(\lambda)$ is decreasing, and

$$x(\lambda) \to \begin{cases} U & \text{as } \lambda \to -\infty, \\ L & \text{as } \lambda \to +\infty. \end{cases}$$

From this we clearly have the following:

Corollary 13.1. *Consider, for $y \in \mathbb{R}$, the equation (in λ)*

$$x(\lambda) = L\frac{e^{-\lambda L}}{e^{-\lambda L} + e^{-\lambda U}} + U\frac{e^{-\lambda U}}{e^{-\lambda L} + e^{-\lambda U}} = y. \tag{13.12}$$

Then for $y \in (L, U)$, there is a unique λ^* such that $x(\lambda^*) = y$. If either $y = L$ or $y = U$, then $x(\lambda) \to L$ (or $x(\lambda) \to U$) as $\lambda \to \infty$ (respectively, $\lambda \to -\infty$).

To finish:

Proof of Theorem 13.1. To verify the claim, note that, for example,

$$\Sigma(\lambda) = \lambda(y - L) + \ln(1 + e^{-\lambda(U-L)}) \to -\infty$$

as $\lambda \to \infty$ and $y < L$. The other case is similar. □

Comment. Note that when $y \in (U, L)$, (13.12) is the first-order condition for λ^* to be a minimizer of $\Sigma(\lambda)$.

14 Review of statistical procedures

Even though we assumed at the outset that the readers of this book are familiar with basic probability theory, we do not assume that they are necessarily familiar with most statistical techniques and concepts that are necessary for the examples analyzed in this volume.

Even though statistical estimation and hypothesis testing are tools that most applied mathematicians are familiar with, we review that material anyway. There are several sections devoted to cluster analysis and the EM method. These are important for the disentangling problems discussed in Chapter 11. There are various sections devoted to different methods to quantify the quality of the maxentropic reconstructions and the different goodness of fit methods and the error measurement. Also, there is a lengthy section devoted to copulas in which a collection of examples is examined.

14.1 Parameter estimation techniques

Here we rapidly review two standard techniques: the maximum likelihood method and the method of moments. This last method is related to the Laplace transform methodology that we proposed for dealing with densities of positive random variables.

14.1.1 Maximum likelihood estimation

Maximum likelihood estimation, also known by its acronym MLE, is used in statistics when we suspect that a given dataset follows some parametric distribution, or that the data is sampled from such a distribution. Is this case only the parameters characterizing the distribution need to be determined. The choice of the family of distributions is in most cases determined by a comparison between the theoretical and the empirical distributions, and/or from some previous knowledge about the underlying process that generated the data.

Suppose we have a sample $X = \{x_1, x_2, \ldots, x_n\}$ of n independent observations from $f(x \mid \theta)$, $x \in X$, with θ being an unknown vector of parameters. We must estimate a $\hat{\theta}$, which is as close as possible to the true value θ, using the sample $\{x_1, x_2, \ldots, x_n\}$.

The maximum likelihood estimation method begins with the definition of a mathematical expression known as the likelihood function, which depends on the data and the unknown parameters. The parameter values that maximize the likelihood function are known as maximum likelihood estimators. The likelihood function is defined as

$$L(\theta; x_1, x_2, \ldots, x_n) = f(x_1 \mid \theta) \cdot f(x_2 \mid \theta) \cdots f(x_n \mid \theta) = \prod_{i=1}^{n} f(x_i \mid \theta).$$

In practice, the logarithm of this function is used to facilitate calculations. Thus, the maximum likelihood estimator $\hat{\theta}$ is the one that maximizes the likelihood function, i. e.,

$$\hat{\theta} = \underset{\theta \in \Theta}{\operatorname{argmax}} L(\theta; x_1, x_2, \ldots, x_n),$$

where Θ stands for the range of the parameters.

To find the extremal points, analytical methods are usually used when the likelihood function is simple, that is, the partial derivatives ($\frac{\partial L(\theta, x)}{\partial \theta}$) are equated to 0 and the resulting system of equations is solved. But we may have to resort to numerical optimization methods. It may also happen that the maximum is not unique.

In R we can use the maximum likelihood method using any of these two commands:
(1) *mle()* is included in the library *stats4* and it requires as one of its parameters the negative likelihood function.
(2) *fitdistr()* is included in the library or package *MASS*.

14.1.2 Method of moments

The simplest version of this method is based on the relationship that may exist between the parameters of a distribution and its moments. It also depends on the possibility of estimating the moments accurately (invoking the law of large numbers). To put it simply, the idea goes as follows: Once the relation between moments and parameters has been established, estimate the sample moments from a large sample X of size n. In this way the parameters of interest may be obtained solving a nonlinear system of equations.

In order to understand the method, suppose that we want to estimate the parameters p and m of a binomial distribution using the method of moments. The relation between the first two moments and the parameters is given by

$$\mu_1 = E[X] = mp \quad \text{and} \quad \mu_2 = E[X^2] = m^2 p^2 + mpq.$$

Now each of the theoretical moments is equated to its corresponding sampling moments, obtaining a system of two equations with two unknowns:

$$mp = \frac{1}{n} \sum_{i=1}^{n} x_i = \overline{X}$$

$$m^2 p^2 + mpq = \frac{1}{n} \sum_{i=1}^{n} x_i^2 = \overline{X^2}.$$

Finally,

$$m = \frac{\overline{X}^2}{\overline{X}^2 + \overline{X} - \overline{X^2}}$$

$$p = \frac{\overline{X}^2 + \overline{X} - \overline{X^2}}{\overline{X}} = \frac{\overline{X}}{m}.$$

Comment. We close this section by mentioning once again that one of the most widely used applications of the maximum entropy method is to solve this type of moment problem.

14.2 Clustering methods

Consider a situation in which we cannot distinguish between two or more subpopulations of risk sources. It is then necessary to separate those subpopulations in order to calculate the underlying distributions as well as the different measures of risk. That can be done using clustering methods, for example K-means, which group observations in such a way that the data within any two groups are similar while data across groups are different. Those similarities and differences can be characterized mathematically in terms of distance metrics, which provides different segmentation solutions.

There are many statistical methods in the literature that are used to separate groups of observations. In practice, one may use various approaches and then select the most robust solution. Here, we discuss two widely used methods named K-means and EM algorithms. Later, we present another advanced methodology called projection pursuit, which is an alternative method to take in account. Other methods that we do not discuss here, but that may be of interest for the reader, are K-medians, hierarchical clustering, and T-SNE.

Before proceeding with the separation of the data in clusters, a few steps are required. It is necessary to decide which variables to use, then decide whether to scale or standardize the data and later perform an initial analysis of the data in order to check for any inconsistencies. This procedure may vary depending on the particular application and its success in similar situations.

14.2.1 K-means

K-means is a well-known partitioning method, and the most popular clustering algorithm used in scientific and business applications. The method was proposed in [51] and improved in [52]. It consists of partitioning the observations of a dataset X into K groups, such that the sum of squares of the observations assigned to the same cluster center is minimal. This means that each observation is assigned to the cluster with the smallest within-group sum of squared errors (SSE):

$$\text{SSE}(k) = \sum_{i=1}^{n} \sum_{j=1}^{p} (x_{ij} - \bar{x}_{kj})^2,$$

where k is the label or identification of the cluster, p is the number of variables used, x_{ij} is the value of the j-th variable for the i-th observation, and $\bar{x}_{kj}$ is the mean of the j-th variable for the k-th cluster.

Briefly, the algorithm starts after choosing a number $k \leq K$ of initial centroids (the mean of a group of points), where K is specified by the user. Then, each point is assigned to the closest centroid, and each collection of points assigned to a centroid is a cluster. The centroid of each cluster is then updated based on the points assigned to the cluster. We repeat the procedure and update the steps until no point changes clusters, or until the centroids remain the same.

14.2.2 EM algorithm

The EM algorithm is a technique widely used to deal with incomplete or missing data. Briefly, this method consists of finding the maximum likelihood estimates of the parameters of a distribution from a given incomplete dataset and it requires two steps: the computation of the expectation (the E-step) and the maximization (M-step) of the likelihood.

The EM algorithm is useful when data values are missing due to limitations in the observation process or due to the nature of the data. Here, we may assume that there is an unobserved variable that indicates the group to which a given datum belongs. At the end of each iteration cycle of the EM method we end up with a probability vector that indicates the likelihood that a data point belongs to a given group. In this way, the point ends up being assigned to the group of higher probability of belonging.

To better describe the algorithm, consider X as the observed data generated by a mixture of H distributions, each having a probability π_h with density function f_h and parameter(s) θ_h such that

$$f(x; \Theta) = \sum_{h=1}^{H} \pi_h f_h(x; \theta_h), \tag{14.1}$$

where $\Theta = \{(\pi_h, \theta_h) : h = 1, \ldots, H\}$ is used to denote the unknown parameters, with $0 \leq \pi_h \leq 1$ and $\sum_{h=1}^{H} \pi_h$ for any number H of groups. The log-likelihood of (14.1) can be written as

$$\ln L(\Theta) = \sum_{i=1}^{n} \log\left(\sum_{h=1}^{H} \pi_h f_h(x_i; \theta_h)\right) \tag{14.2}$$

for any given data vector $X = \{x_i : i = 1, \ldots, n\}$. Introducing the variable Z, which represents the unobserved data, the log-likelihood of the augmented or completed data $\{X, Z\}$ is

$$\ln L_c = \ln L(\Theta) = \sum_{i=1}^{n} \sum_{h=1}^{H} z_{ih} \log(\pi_h f_h(x_i; \theta_h)). \tag{14.3}$$

The E-step consists of the computation of the expected value of the log-likelihood with respect to the distribution of the unobserved variable. The result is denoted by Q:

$$Q(\Theta; \Theta^{(m)}) = E_{\Theta^{(m)}}(\ln L_c \mid X) = \sum_{i=1}^{n} \sum_{h=1}^{H} \tau_{ih}^{(m)} \log(\pi_h f_h(x_i; \theta_h)), \quad (14.4)$$

where $\Theta^{(m)}$ represents the parameters to be estimated at the m-th iteration. In the M-step the estimated parameters that maximize Q are updated. Both steps are repeated until convergence is clear. Note that the relationship between z and τ is contained in

$$\tau_{ih} = E[z_{hi} \mid x_j, \hat{\theta}_1, \ldots, \hat{\theta}_H; \hat{\pi}_1, \ldots \hat{\pi}_H]. \quad (14.5)$$

At the end of the process the parameters $\{\hat{\theta}_1, \ldots, \hat{\theta}_H; \hat{\pi}_1, \ldots \hat{\pi}_H\}$ are obtained. These are the maximizers of the expected log-likelihood of a mixture of distributions of H components (see [68]):

$$\hat{\Theta} = \operatorname*{argmax}_{\Theta} \log L(\Theta).$$

A handicap of the EM and K-means algorithms is that the number of groups must be specified in advance. There are some criteria that can be used to overcome this limitation, as for example those denoted by their acronyms AIC, AIC3, BIC and ICL-BIC, and thoroughly described in Section 14.3 of this chapter. These metrics are computed for different values of the number of groups H, and the number yielding the lowest value of the metric is chosen as the number of clusters.

14.2.3 EM algorithm for linear and nonlinear patterns

Let us consider a few concrete applications of the EM method to disentangle linear and nonlinear regressions. In Chapter 11 we used this methodology to separate linear regressions of the type $\hat{y} = \hat{\beta}_o + \sum_i \hat{\beta}_i x$, which means $\hat{\mu} = E(\hat{y}) = \hat{\beta}_o + \sum_i \hat{\beta}_i x$, parametrized by $\hat{\theta}_h = \{\hat{\beta}_{oh}, \hat{\beta}_{1h} \ldots, \hat{\beta}_{mh}, \hat{\sigma}_h^2\}$. In that case the set of parameters to estimate are $\{\hat{\beta}_{oh}, \hat{\beta}_{1h} \ldots, \hat{\beta}_{mh}, \hat{\sigma}_h^2; \hat{\pi}_h\}$ for $h = 1, \ldots, H$ with H the number of groups. To determine the exact positioning of the lines, it is easy modify the EM algorithm to differentiate between linear and nonlinear regressions.

In the examples described below, we compare the modified EM algorithm with the algorithm *Mclust* from the library mclust of R, in order to see the differences between the original method and the modified EM method for the cases of linear and nonlinear patterns.

Example 1. In this example we consider a case in which there five groups that appear to follow linear patterns. Looking at Figure 14.1 is it not clear at first sight what the number of groups is or which points correspond to a particular cluster.

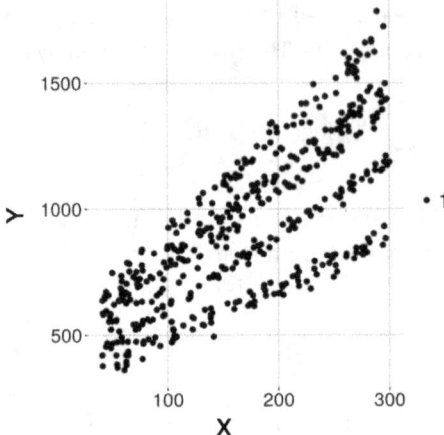

Figure 14.1: Linear patterns.

In Figure 14.2 we can see the differences between the results produced by the Mclust algorithm of R (left panel) and the modified algorithm (right panel). Here, it is clear that the modified algorithm performs better at finding linear patterns as was expected. Clearly, this methodology has a certain error, but in general works well.

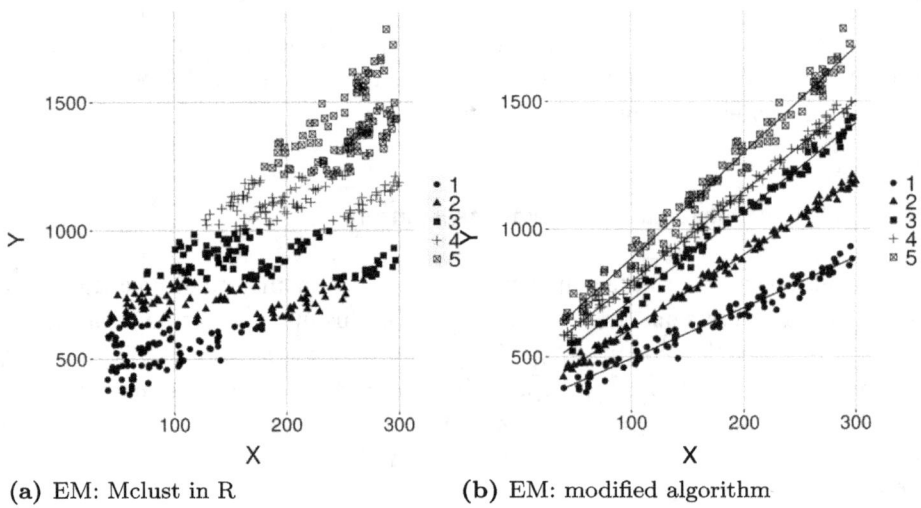

(a) EM: Mclust in R (b) EM: modified algorithm

Figure 14.2: Result of applying the EM algorithm.

Example 2. Now, consider a nonlinear case as in Figure 14.3.

In Figure 14.4 we can see the differences between the output of the Mclust algorithm of R and that of the modified algorithm for this example; again the modified EM algorithm performs better.

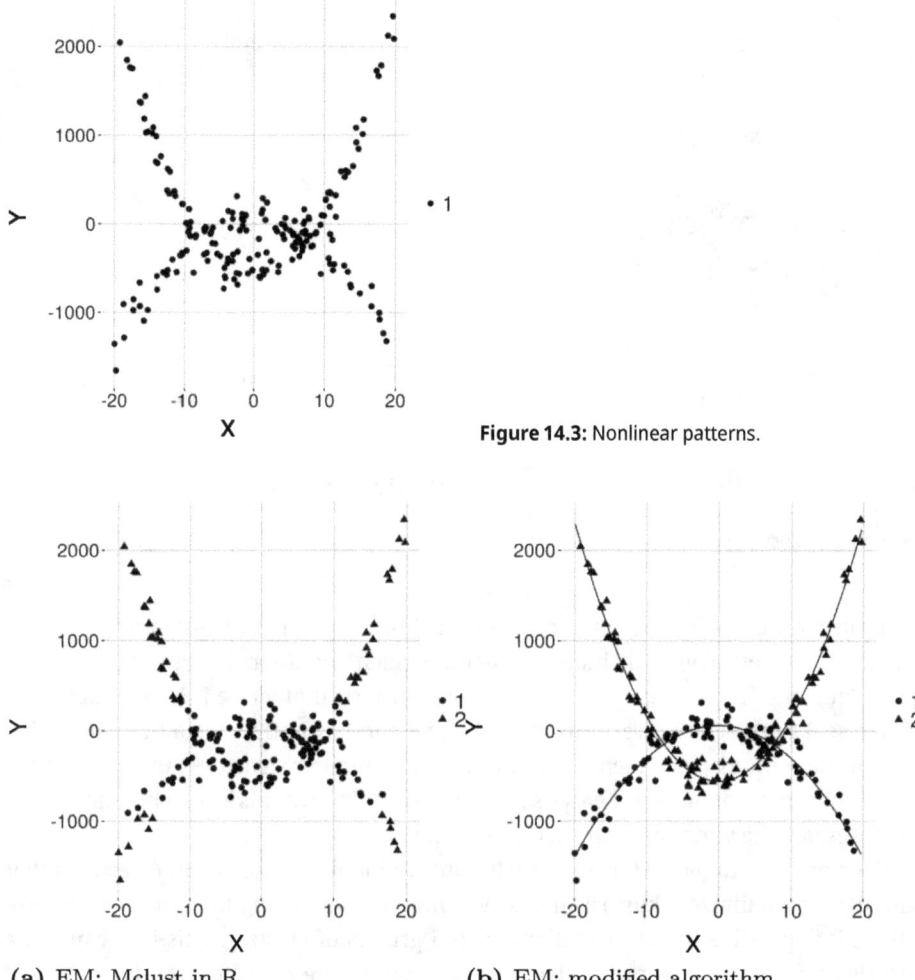

Figure 14.3: Nonlinear patterns.

(a) EM: Mclust in R (b) EM: modified algorithm

Figure 14.4: Result of applying the EM algorithm.

14.2.4 Exploratory projection pursuit techniques

Imagine you have a cloud of points that you know belong to two or more distributions with unknown mean and variance as can be seen in the left panel of Figure 14.5. One way to find such distributions is to project the data onto one of the coordinate axes, rotating the data points a degrees, so that the data is grouped into two or more distributions with the greatest possible distance between them, as is shown in the right panel of Figure 14.5. Then, the search for the a direction that produces the maximum gap or separation between the projected data is what it is known as 'projection pursuit'.

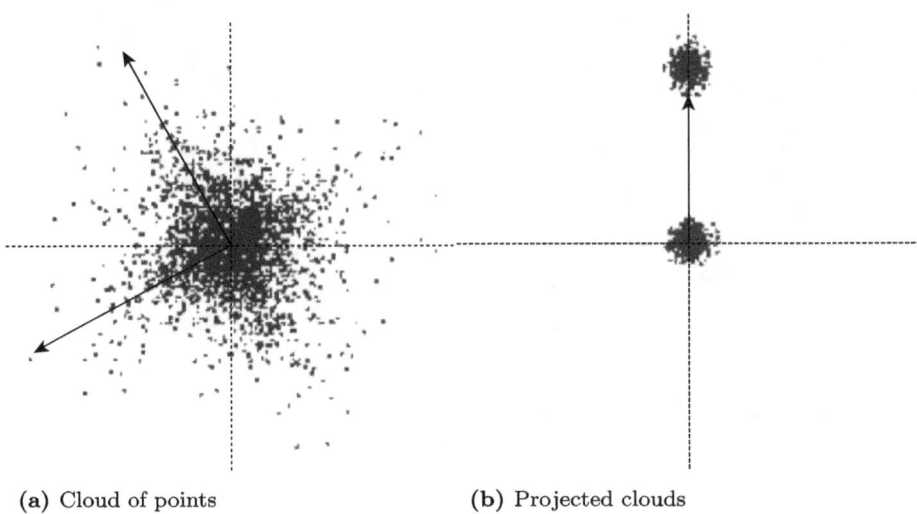

(a) Cloud of points (b) Projected clouds

Figure 14.5: Projection pursuit.

The projection pursuit technique is very useful to complement exploratory analysis, since it aims to reveal visually hidden structures, such as outliers in high or low dimensional data, and helps to assess the number of clusters, if any (see [18]). Basically, this method seeks projection directions (or weight vectors) by optimizing a function called the projection index $I(\cdot)$, which serves to capture nonlinear structures in a big volume of data. Principal component analysis (PCA) is a technique of this kind where the index $I(\cdot)$ represents the variance of the projected data [89].

Peña and Prieto [80, 81] use this technique and look for projection directions that minimize or maximize a kurtosis index. Viewing the data along those univariate projections it is possible to detect outliers or find groups of clusters. This is because the kurtosis coefficient is inversely related to bimodality. The presence of outliers would thus cause heavy tails and a large kurtosis coefficient, while clusters or larger number of outliers may introduce multimodality and the kurtosis would decrease (see [98]). Thus, the detection of clusters is achieved by minimizing this function while maximizing outlier detection. Additionally, the presence of significant groups or clusters is controlled through the spacing of the projected points and a readjustment procedure, based on the Mahalanobis distance to verify the quality of the resulting groups.

The advantage of this method is that no preliminary knowledge of the number of groups is required. However, this method should be used with caution, since it tends to produce a relatively large number of false positives in some cases. Alternative methods of clustering methods may then be used to complement and verify the results. Consult with [80, 81] or [98].

14.3 Approaches to select the number of clusters

Let us now address one of the most controversial issues in clustering and describe some statistical criteria that can help in this task. Several approaches have been proposed to determine the correct number of clusters H, a value that it is necessary for the implementation of algorithms like K-means and EM. Determining the true number of clusters or groups present in a dataset is often difficult, and there is no one true method that works best in all possible cases. In the end, it is always necessary to use the best judgment of the analyst. Among the methods we consider the following approaches: (1) elbow method, (2) information criterion approach, (3) negentropy, and (4) projection pursuit, which we already described in the previous section.

14.3.1 Elbow method

This is a visual method; specifically the method consists of calculating the within-group sum of squares of errors (WSSE) for each group or partition given after applying the K-means, EM or any other procedure and looking for an elbow (value where WSSE drops dramatically) in the resulting curve. This method can be ambiguous, because is not always clear what the best answer is.

14.3.2 Information criteria approach

It is customary to utilize a multiplicity of information criteria like the information criterion (AIC) of [3], modified AIC criterion (AIC3) of [14], Bayesian information criterion (BIC) of [86] and the integrated classification likelihood BIC-type approximation (ICL-BIC) of [11] to select the best model. In order to determine the number of groups present in the mixture, the clustering procedure must be executed for several values of classes or groups. These measures employ the log-likelihood at convergence of a specific clustering method for each number of classes or groups (H), sample size (n) and number of parameters. These measures address goodness of fit of a specific method, where the lowest value is best. Each of these measures penalizes models with more parameters, promoting parsimony; conversely, the BIC (Bayesian information criterion) penalizes increases in the sample size.

In summary, this methodology yields a table that contains the metrics mentioned before for each possible value. The criteria to select the number of clusters will be that with the lowest value in the majority of the criteria. The Akaike information criterion [4] takes the form

$$AIC = -2\log lik + 2H.$$

Many authors observed that AIC tends to overestimate the correct number of components. Thus, in order to correct that limitation, this criteria leads to the modified AIC criterion or AIC3:

$$\text{AIC3} = -2\,\text{loglik} + 3H.$$

The Bayesian information criterion (BIC) penalizes complex models more heavily than AIC, whose penalty term H does not depend on the sample size. The BIC is given by

$$\text{BIC} = -2\,\text{loglik} + H\log(n),$$

where loglik is the likelihood of the mixture model.

Finally, the integrated classification likelihood criterion (ICL-BIC) includes an entropy measure to improve its ability to estimate the correct number of components:

$$\text{ICL} = \text{BIC} + 2 * \text{ENT},$$

where ENT is the entropy measure, calculated as $-\sum_{i=1}^{n}\sum_{l=1}^{H} p_{il} \cdot \ln(p_{il}) \geq 0$, where the classification matrix p_{il} gives the posterior probabilities that each data point belongs to a specific group; this may be obtained by Bayes' theorem.

14.3.3 Negentropy

Another measure of interest is negentropy or negative entropy, which is a measure varying from 0 to 1, and indicates how well discriminated or separated the classes seem to be, based on the posterior probabilities. If the model's assignment mechanism is indistinguishable from random assignment, the negentropy will therefore zero; on the contrary if the classes are well distinguished, the negentropy will be nearer to one ([82] and [90]). The negentropy is expressed as

$$\text{Negentropy} = 1 - \frac{\sum_{i=1}^{n}\sum_{l=1}^{H} -p_{il} \cdot \ln(p_{il})}{n \cdot \ln(H)},$$

where H is the number of groups, n is the number of elements in the data and p_{il} is the posterior probability that the element i is in group l; this can be obtained by Bayes' theorem.

14.4 Validation methods for density estimations

Once a density function has been obtained, it is necessary to test whether it is consistent with the data. The evaluation process is inherently a statistical problem, which involves exploring, describing, and making inferences about datasets containing observed and estimated values. Exploratory tests to asses the quality of a reconstruction include visual comparisons, through the use of graphical tools like reliability and calibration plots, which measures the agreement between the estimation and the observed data, statistical tests and error measurement.

In this section we describe several of the tests that have been proposed to assess the quality of a probability distributions estimated from available data. We applied them to evaluating the quality of the output of the procedures in the numerical examples that we considered in the previous chapters.

14.4.1 Goodness of fit test for discrete variables

Chi-squared test
This is a simple test that is used to verify whether a given discrete distribution fits the data appropriately or not. Here, the test statistic is calculated as

$$\chi_o^2 = \sum_{j=1}^{k} \frac{(E_j - O_j)^2}{E_j},$$

where k is the number of categories, discrete values or partitions, the value $E_j = n\hat{p}_j$ is the expected number of observations for a specific discrete value or category (assuming that the hypothesized model is true) and $O_j = np_{nj}$ is the observed number of values in a given partition or category.

The critical value for this test comes from the chi-square distribution with degrees of freedom equal to the number of terms is the sum k minus the number of estimated parameters p minus 1. The null hypothesis H_o of no difference between the data population and the stated model has to be rejected at the chosen significance level α of 0.1, 0.05 or 0.01, whenever $\chi_o^2 > \chi_{\alpha,k-1-p}^2$ or if $p_{\text{value}} < \alpha$. More detailed information may be seen in [78].

14.4.2 Goodness of fit test for continuous variables

Probability integral transform (PIT) based analysis
This methodology relies on the fact that $F_S(S)$ is uniformly distributed in $[0, 1]$. Thus, if the s_j are the sample points examining the collections $F_S^*(s_j)$, for uniformity, or deviation thereof, we decide whether reconstruction may have failed to capture some aspect of the underlying data generation process.

To test for uniformity and independence using the PIT test, a visual inspection of a PIT histogram and autocorrelation plots are used along with additional tests like the KS-test, the Anderson–Darling test, and the Cramér–von Mises test, [92]. Additionally, we may also consider the Berkowitz back test approach, which consists of taking the inverse normal transformation of the PIT and then applying a joint test for normality and independence, which is sometimes combined with the normality test of Jarque–Bera. Let us briefly describe some of these tests now.

Kolmogorov–Smirnov test
The Kolmogorov–Smirnov test is a test of uniformity based on the differences of fit between the empirical distribution function (EDF) and the estimated (reconstructed) distribution function. The quality of fit is measured by calculating the largest absolute observed distance between them, as

$$D_n = \sup_s |F_n(s_j) - F_S^*(s_j)|,$$

where n is the number of data points, $\{s_j \mid j = 1, \ldots, n\}$ are the sample data points of the total losses S, $F_n(\cdot)$ is the (cumulative) empirical distribution function, and $F_S^*(\cdot)$ is the maxentropic (cumulative) distribution function. The statistic to be used to perform the test is $\sqrt{n}D_n$. A problem with this test is that the KS statistic depends on the maximum difference without considering the whole estimated distribution. This is important when the differences between distributions are suspected to occur only at the upper or lower end of their range. This may be particularly problematic in small samples. Besides, little is known about the impact of the departures from independence on D_n. That is, if we are not sure about the independence of the sample, we would not be sure of the meaning of the results of the test. There exist other EDF tests, which in most situations are more effective than the simple Kolmogorov–Smirnov test. Further details about the K-S test can be seen in [88] or in [85].

Anderson–Darling test
This is a more sophisticated version of the KS approach, which emphasizes more the tails of the distribution than the KS test, and is based on the quadratic difference between $F_n(s)$ and $F_S^*(s)$. Here the Anderson–Darling (AD) statistic is computed as follows:

$$A_n^2 = n \int_{-\infty}^{\infty} |F_n(s) - F_S^*(s)|^2 \Psi(s) f^*(s) ds,$$

where $\Psi(s) = \frac{1}{F_S^*(s)(1-F_S^*(s))}$ is a weight function, n is the number of data points, $\{s_j \mid j = 1, \ldots, n\}$ is the observed (simulated in our case) total loss S sample, $F_n(\cdot)$ is the empirical (cumulative) distribution function and $F_S^*(\cdot)$ is the (cumulative) maxentropic distribution function.

When $\Psi(s) = 1$ the AD statistic reduces to the statistic that is today known as the Cramér–von Mises statistic. The AD statistic behaves similarly to the Cramér–von Mises statistic, but is more powerful for testing whether $F_S^*(s)$ departs from the true distribution in the tails, especially when there appear to be many outlying values. For goodness of fit testing, departure in the tails is often important to detect, and A_n^2 is the recommended statistic. For this, see [67] and [94].

Berkowitz back test

Berkowitz [10] proposed the transformation $z_n = \Phi^{-1}(\int_{-\infty}^{S_n} f_S^*(s)ds) = \Phi^{-1}(F(s_n))$, to make the data IID standard normal under the null hypothesis. This allows one to make use of a powerful battery of available normality tests, instead of relying on uniformity tests. Besides that, the Berkowitz back test provides a joint test of normality and independence. See [19] for more details.

It is usually convenient to supplement the Berkowitz test with at least one additional test for normality, for example the Jarque–Bera test. This extra test ensures that we test for the predicted normal distribution and not just for the predicted values of the parameters ρ, μ and σ.

Jarque–Bera test

The standard Jarque–Bera (JB) test is a test of normality. This test uses the empirical skewness coefficient and the empirical kurtosis in a statistic to test deviations from the normal distribution.

Unfortunately, the standard JB statistic is very sensitive to extreme observations, due to the fact that the empirical moments are known to be very sensitive to outliers, and that the sample variance is more affected by outliers than the mean, disturbing the estimations of the sample skewness and kurtosis. To solve the problem a robust modification of the Jarque–Bera test was proposed in [40], which utilizes the robust standard deviation (namely the average absolute deviation from the median (MAAD)) to estimate a more robust kurtosis and skewness from the sample. For this, see [40] and consider [85].

Correlograms

Tests like KS and AD do not prove independence, so to asses whether the probability integral transformation (PIT) of the data is IID, we use a graphical tool, the correlogram, which helps in the detection of particular dependence patterns and can provide information about the deficiencies of the density reconstructions [29].

14.4.3 A note about goodness of fit tests

To close this section, let us note that the KS, AD, and the Cramér–von Mises tests are the most used goodness of fit tests in the context of operational risk. These tests are more powerful than the chi-squared test, in the sense that they are less likely to reject a null hypothesis when the null hypothesis is true. However, these tests are dependent on the sample size, so it is always possible to reject a null hypothesis with a large enough sample, even if the true difference is trivially small. The solution here is estimate the effect size (magnitude of the differences) that we want to test and calculate a measure of its precision (confidence intervals and power) to guarantee that there is a real difference between the distributions that we are comparing [2].

14.4.4 Visual comparisons

Let us now examine some 'visual comparison' tests. That is, let us go a bit beyond the 'it looks nice' approach.

Reliability diagram or PP-plots
This plot serves to determine the quality of a fit by the proximity of the fitted curve to the diagonal. The closer to the diagonal the better the approximation; deviations from the diagonal give the conditional bias. Additionally, this plot could indicate the existence of problems like overfitting when the fitted curve lies below the diagonal line, and underfitting when the fitted curve lies above the line (see [55]).

Marginal calibration plot
The calibration plot is based in the idea that a system is marginally calibrated when its estimations and observations have the same (or nearly the same) marginal distribution. Then, a graphical device consists of making a plot of $F_S^*(s_j) - F_n(s_j)$ versus s_j. Under the hypothesis of marginal calibration, we expect minor fluctuations about zero. The same information may be visualized in terms of the quantiles $Q(F_S^*(\cdot), q) - Q(F_n(\cdot), q), q \in (0, 1)$ of the functions $F_S^*(\cdot)$ and $F_n(\cdot)$; see [42].

14.4.5 Error measurement

In order to analyze the quality of the numerical results, we might use measures of error, such as L1, L2, MAE and RMSE distances. The first two measure the distance between the fitted density and the histogram of the data sample. The second two consider the cumulative distribution function (CDF) to calculate the difference between the fitted CDF and the sample CDF.

Distances in L_1 and L_2 norms
This assessment error is based on the evaluation of the distances between the histogram and the fitted density by means of

$$L_1 = \sum_{k=0}^{G-1} \int_{b_k}^{b_{k+1}} |f_S^*(s) - f_n(s)| ds + \int_{b_G}^{\infty} |f_S^*(s)| ds$$

$$L_2 = \sqrt{\sum_{k=0}^{G-1} \int_{b_k}^{b_{k+1}} (f_S^*(s) - f_n(s))^2 ds + \int_{b_G}^{\infty} (f_S^*(s))^2 ds},$$

where b_k and b_{k+1} are the boundary points of the bins of the histogram, G is the number of bins, f_S^* is the reconstructed density and f_n is the empirical density obtained from the data (i. e., frequency in the bin k/size of the dataset). This measure has the disadvantage of depending on the location and the number of bins of the histogram.

MAE and RMSE

To overcome the bin dependency we can compute the mean absolute error (MAE) and the root mean squared error (RMSE), which measures the distance between the cumulative distribution functions of the empirical and the CDF of the reconstructed densities. These measures are computed as follows:

$$\text{MAE} = \frac{1}{n} \sum_{j=1}^{n} |F_S^*(s_j) - F_n(s_j)|$$

$$\text{RMSE} = \sqrt{\frac{1}{n} \sum_{j=1}^{n} (F_S^*(s_j) - F_n(s_j))^2}.$$

Note that the computations do not depend on the choice of bins to create the histogram, but only on the sample data; see [54].

The measure RMSE is more sensitive to outliers because it gives a relatively high weight to large errors. So, the greater the difference between MAE and RMSE, the greater the variance of the individual errors in the sample.

14.5 Beware of overfitting

One important question when we are in the process of modeling is: Which is the best criteria to use when we have to select a model among a set of competing models? A short answer is that a model should be selected based on its generalizability, by which we mean the model's ability to fit current data and also perform well on new data. This is important in order to avoid the problem of overfitting, which occurs when the model gives better results for the set than for other datasets that come from the same population.

A good fit is a necessary, but not a sufficient condition to ensure that an approximation correctly captures the underlying distribution. This is because a model can achieve a superior fit than its competitors for reasons that may have nothing to do with the model's faithfulness to the underlying data. So, using a set of values that come from the same underlying population, but that was not used in the modeling process, is a good additional element to test the quality of the results.

14.6 Copulas

A copula is a tool used to create statistical dependence among a collection of random variables whose individual distribution functions are known. Its main purpose is to describe a possible interrelationship between several random variables. The idea of a function that characterizes the dependence structure between different random variables goes back the work of Hoeffding in the mid 1940s, although it was Sklar in 1959 who defined and established the name of copula ([76] and [32]). Copulas are a very useful not only for modeling, but also for the estimation or simulation of random variables. As we have made extensive use of this technique when working out the problem of aggregating risk with several levels of aggregation, we shall devote the rest of this chapter to the underlying examples.

Definition

A copula is a joint distribution function of a collection $\{U_1, U_2, \ldots, U_n\}$ of random variables taking values in $[0, 1]$ having uniform one dimensional marginals. Formally, it can be expressed as:

$$C(u_1, u_2, \ldots, u_n) = P[U_1 \leq u_1, U_2 \leq u_2, \ldots, U_n \leq u_n]. \tag{14.6}$$

Properties

Since a copula is a d-dimensional distribution function, it satisfies the following:
1. $C: [0, 1]^d \to [0, 1]$.
2. For any $u_j \in [0, 1]$ on $j \in \{1, \ldots, d\}$, it holds that $C(0, 0, \ldots, u_j, \ldots, 0) = 0$.
3. For any $u_j \in [0, 1]$ on $j \in \{1, \ldots, d\}$, it holds that $C(1, 1, \ldots, u_j, \ldots, 1) = u_j$.
4. C is d-ascending, i.e., for all $(u_1, \ldots, u_n), (v_1, \ldots, v_n) \in [0, 1]^d$ with $u_j \leq v_j$ we have

$$\sum_{i_1=1}^{2} \cdots \sum_{i_d=1}^{2} (-1)^{i_1 + \cdots + i_d} C(g_{1i_1}, \ldots, g_{di_d}) \geq 0, \tag{14.7}$$

where $g_{j1} = u_j$ and $g_{j2} = v_j$ for all $j \in \{1, \ldots, d\}$.

Conversely, a function that satisfies these properties is a copula. Clearly, it is natural to think that each distribution function in $(R)^d$ determines a copula. Reciprocally, if we start from a copula along with some marginal distributions and combine them appropriately we will obtain a multivariate distribution function. This is the content of the following result.

Sklar's theorem

Let F be a joint distribution function with marginal $F_1, \ldots, F_d$. Then there is a copula C such that for all $x_1, \ldots, x_d \in \mathbb{R}$

$$F(x_1, \ldots, x_d) = C(F_1(x_1), \ldots, F_d(x_d)).$$

If the F_i are continuous, then C is unique. Conversely, if $F_1, \ldots, F_d$ are distribution functions, then the function defined by $F(x_1, \ldots, x_d) = C(F_1(x_1), \ldots, F_d(x_d))$ is a joint distribution function with marginals $F_1, \ldots, F_d$.

This is an important theorem that states that any multivariate distribution admits a representation through a copula function. In addition, if the marginal distributions are continuous, the copula function is unique. Reciprocally, it asserts that given any copula function and a collection of marginal distributions, we obtain a multivariate distribution. This fact is what has made the use of copulas an important tool to generate dependence among random variables with given marginals.

14.6.1 Examples of copulas

Let us examine a few examples. Many of them were used in the previous chapters to generate dependence among the variables describing partial risks. We direct the reader to [32], [76] and to the scattered references for more on the properties of these copulas and further examples.

Maximum copula

This is named this way because it corresponds to the case where the negative maximum dependence occurs. This copula has the following form:

$$W(u_1, u_2, \ldots, u_n) = \max(u_1 + u_2 + \cdots + u_n - n + 1, 0). \tag{14.8}$$

Minimum copula

This is the case where the positive maximum dependence occurs. This copula has the following form:

$$M(u_1, u_2, \ldots, u_n) = \min(u_1, u_2, \ldots, u_n). \tag{14.9}$$

Equations (14.8) and (14.9) are part of the well-known Fréchet inequality for copulas, which states that:

$$W(u_1, u_2, \ldots, u_n) \leq C(u_1, u_2, \ldots, u_n) \leq M(u_1, u_2, \ldots, u_n). \tag{14.10}$$

Independent copula

This clearly describes the case in which there is no dependence between variables:

$$C(u_1, u_2, \ldots, u_n) = \prod_{i=1}^{n} u_i. \tag{14.11}$$

Let us now consider some elliptical copulas, so named because they are derived from elliptical distributions. The advantage of these copulas is that you can specify different levels of correlation between the marginal distributions. The disadvantages are that the elliptic copulas have complicated expressions and are restricted to a radial symmetry [66]. The two most important copulas in this family are the normal copula (or Gaussian) and t-Student copula (or simply t-copula), which are derived from the multivariate distribution functions that have these same names.

Gaussian copula

The copula corresponding to the normal distribution, n-dimensional with linear correlation matrix Σ, is:

$$C_\Sigma^{Ga}(u) = \Phi_\Sigma^n(\Phi^{-1}(u_1),\ldots,\Phi^{-1}(u_n)), \tag{14.12}$$

where Φ_Σ^n denotes the joint distribution function of the standard normal distribution n-dimensional, with linear correlation matrix Σ and Φ^{-1} the inverse distribution function of the univariate normal distribution [66]. For example, for the bivariate case, equation (14.12) is rewritten as:

$$C_\Sigma^{Ga}(u_1, u_2) = \int_{-\infty}^{\Phi^{-1}(u_1)} \int_{-\infty}^{\Phi^{-1}(u_2)} \frac{1}{(2\pi(1-\Sigma^2))^{1/2}} \cdot \frac{s^2 - 2\Sigma st + t^2}{2(1-\Sigma^2)} ds\,dt. \tag{14.13}$$

In Figure 14.6 we show the dependency structure imposed by the Gaussian copula on two simulated normal marginal distributions, each of which has a total of 2000 data points. These two random variables are named X_1 and X_2. Each panel in the figure represents the copula for different values of correlation between the marginal ones. These correlations are −0.8, 0.0, 0.5 and 0.8 respectively.

It is clear in Figure 14.6 that most of the observations are concentrated at the center of the distribution. The figure shows the clear elliptical symmetry of the Gaussian copula.

t-Student copula

This copula is related to the n-dimensional t-Student distribution, with v degrees of freedom and linear correlation matrix Σ:

$$C_{v,\Sigma}^t(u) = \Phi_{v,\Sigma}^n(t_v^{-1}(u_1),\ldots,t_v^{-1}(u_n)), \tag{14.14}$$

where $\Phi_{v,\Sigma}^n$ denotes the joint distribution of an n-dimensional t-Student distribution, with linear correlation matrix Σ, v degrees of freedom and t_v^{-1} the univariate, inverse t-Student distribution, [66]. In the bivariate case we have:

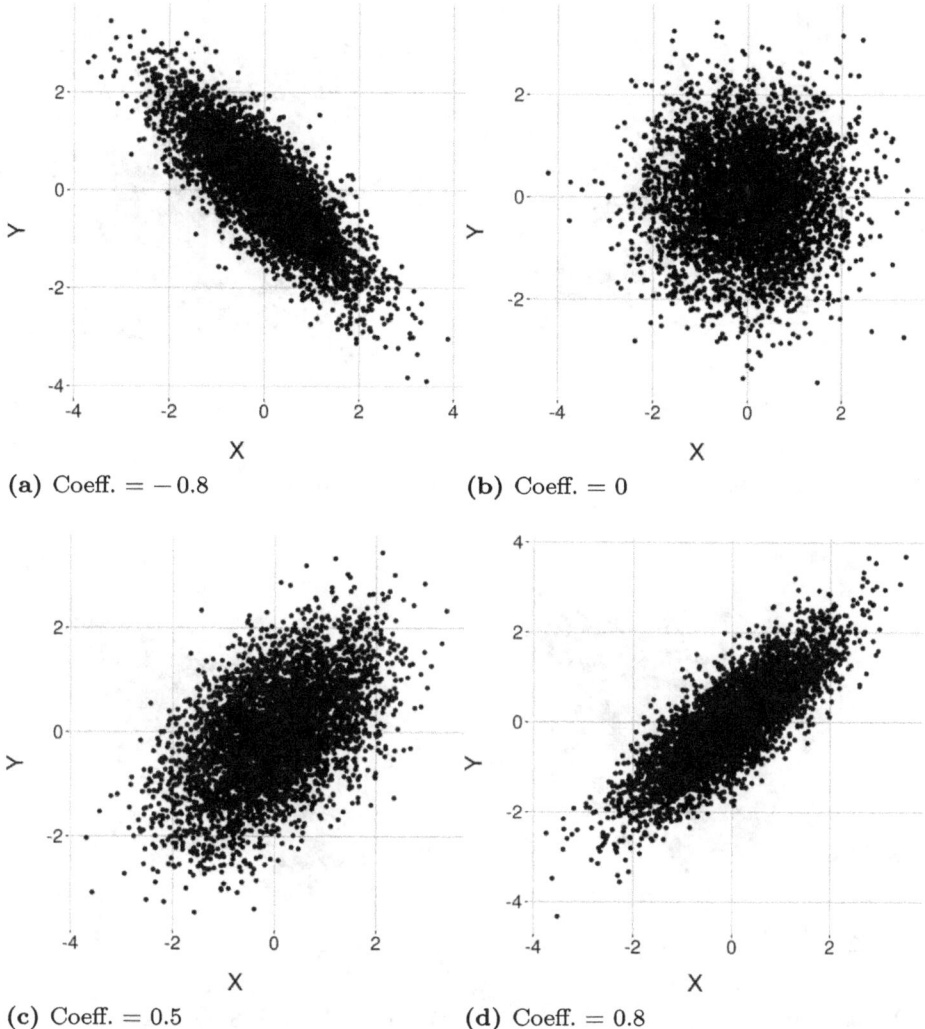

Figure 14.6: Gaussian copulas, bivariate case.

$$C^t_{\nu,\Sigma}(u_1, u_2) = \int_{-\infty}^{t_\nu^{-1}(u_1)} \int_{-\infty}^{t_\nu^{-1}(u_2)} \frac{1}{(2\pi(1-\Sigma^2))^{1/2}} \cdot \left(1 + \frac{s^2 - 2\Sigma st + t^2}{\nu(1-\Sigma^2)}\right)^{-(\nu+2)/2} ds\,dt. \quad (14.15)$$

In Figures 14.7 and 14.8 we show the results of applying a t-copula to two random variables of 2000 points each, with different correlation coefficients, where it is clear that the dependence relation is stronger at the extremes.

In Figure 14.7 we show a t-copula or copula t-Student with fixed degrees of freedom ν and different correlation values, whereas in Figure 14.8 the correlation value is fixed and the number of degrees of freedom (ν) are different.

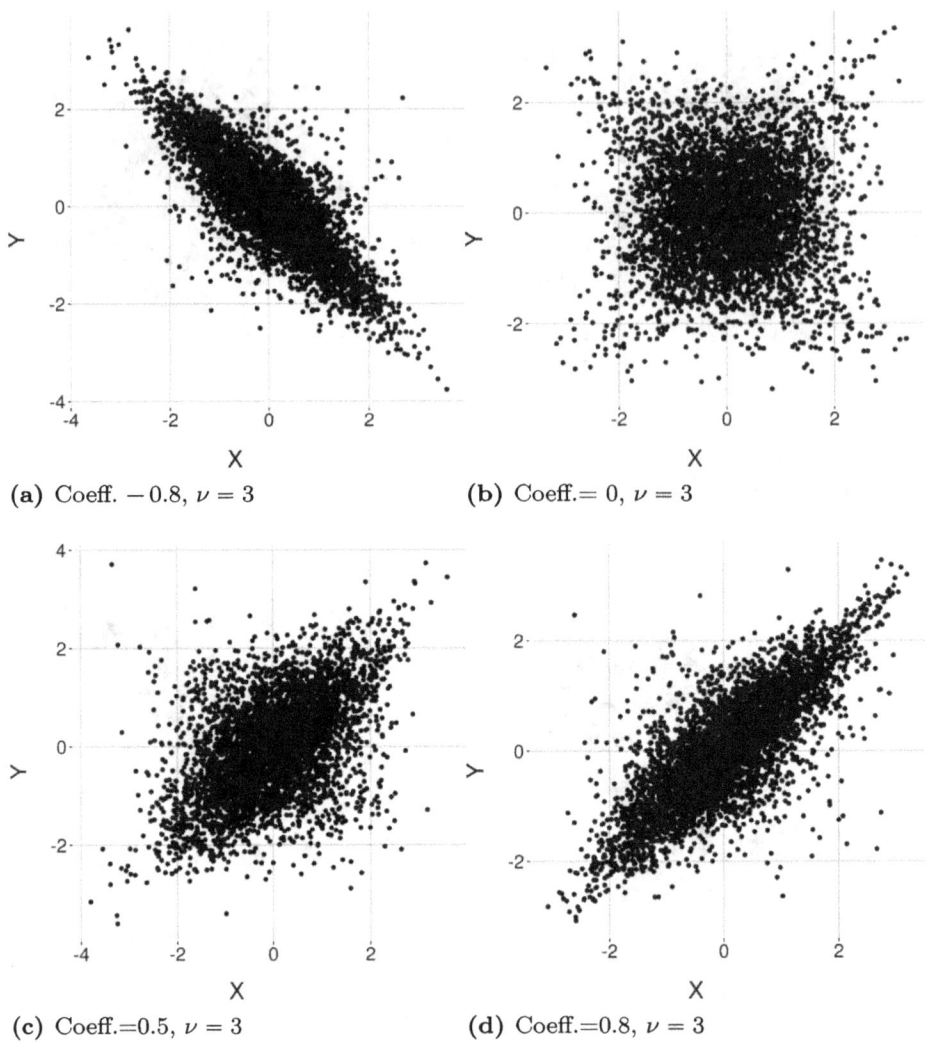

Figure 14.7: *t*-copula, bivariate case with three degrees of freedom.

Figure 14.7 shows how the observations are concentrated at the center of the distribution, which seems to be more dispersed for correlation values lower than 0.5 (and values greater than −0.5) and less dispersed for values closer to 1 (or −1), without losing the radial symmetry that characterizes this type of distributions.

In Figure 14.8 it is observed that larger values of ν (for example $\nu = 1000$) make the copula approximate the Gaussian distribution. In contrast, small values of ν (for example $\nu = 1$ and $\nu = 2$) increase the dependence at the ends of the distribution. Note that in the case when $\nu = 1$, we observe a wing effect in the distribution, an effect that

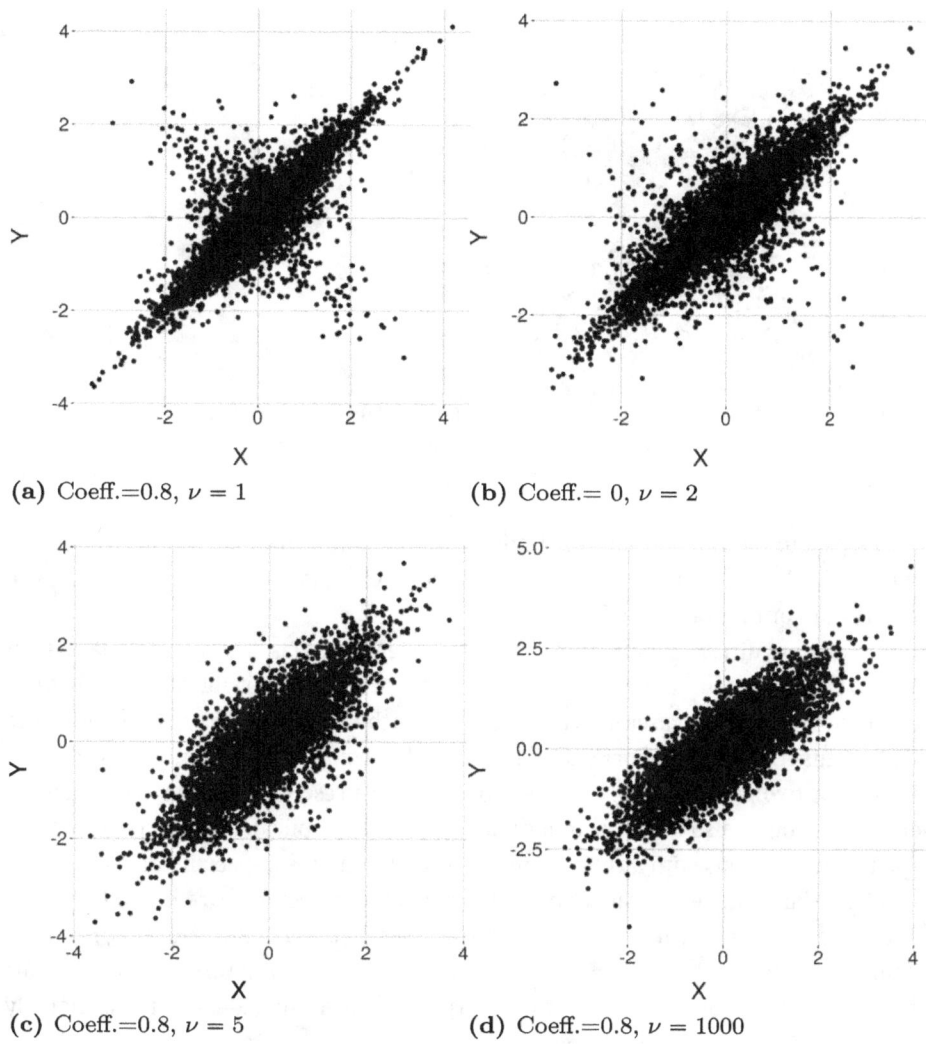

Figure 14.8: *t*-copula, bivariate case with different degrees of freedom.

may not be desired in the case of a simulation. This effect can be observed more clearly in Figure 14.9.

The relation of dependence on the extremes exists and tends to zero as the degrees of freedom tend to infinity. This means that as degrees of freedom increase, the *t*-copula approaches a Gaussian copula.

Despite their complicated expressions elliptic copulas are easy to implement, but they do have some limitations such as that their dependence structure is symmetric. Depending on the application and the data with which you want to work, this feature may not be convenient. However, another advantage of the Gaussian and Student *t*-copulas

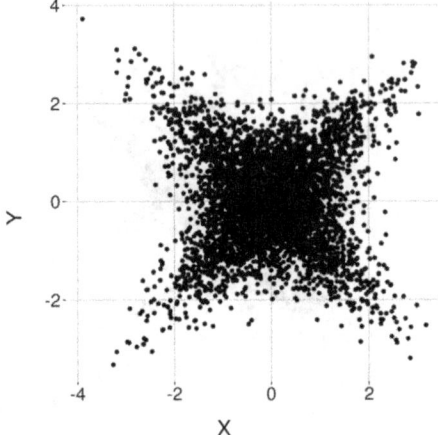

Figure 14.9: Undesired effect for the *t*-copula function.

with respect to the distribution functions from which they originate is that from these copulas it is possible to use random variables that follow marginal distributions and are not Gaussian or *t*-Student (see [66]).

Let us now direct our attention to Archimedean copulas. There are situations in which it is necessary to capture dependence between random variables with some form of asymmetry. For these cases it is convenient to use these copulas. They comprise a large family of functions. The majority of the copulas that belong to this family are functions with one or two parameters. This fact allows us to represent different types of dependence easily, but it also imposes limitations, since it is complicated to describe complex dependency relations with a reduced number of parameters, especially in high dimensions [93]. Unlike the elliptical copulas, the Archimedean copulas are not obtained directly from multivariate distributions and the Sklar theorem. Because of this, much attention is needed on how copulas of this kind can be built. For this consider [93] for example. Next we consider three of the most relevant Archimedean copulas, namely the Gumbel, Clayton, and Frank copulas.

These copulas have the form:

$$C(u_1, u_2, \ldots, u_n) = \Phi^{-1}(\Phi(u_1) + \Phi(u_2) + \cdots + \Phi(u_n)), \qquad (14.16)$$

where Φ is a decreasing function that maps $[0,1]$ and $[0,\infty)$ [66].

Gumbel copula

It is an asymmetric copula that presents dependence only at the upper or lower tail of the distribution. This copula is given by the following expression:

$$C_\theta^{Gu}(u_1, u_2, \ldots, u_n) = \exp(-(-\ln u_1)^\theta + (-\ln u_2)^\theta + \cdots + (-\ln u_n)^\theta)^{1/\theta}, \qquad (14.17)$$

where $\Phi_{Gu}(u) = (-\ln u)^\theta$ with $u = \{u_1, u_2, \ldots, u_n\}$ (see equation (14.16) with $\theta \in [1, \infty)$. Notice that:
- For $\theta = 1$ we obtain the independent copula.
- For $\theta \longrightarrow +\infty$ the expression is reduced to $M(u_1, u_2, \ldots, u_n) = \min(u_1, u_2, \ldots, u_n)$.

Gumbel's copula is an example of an asymmetric copula that has great dependence on the upper end, as can be seen in Figure 14.10, considering different values for the association parameter.

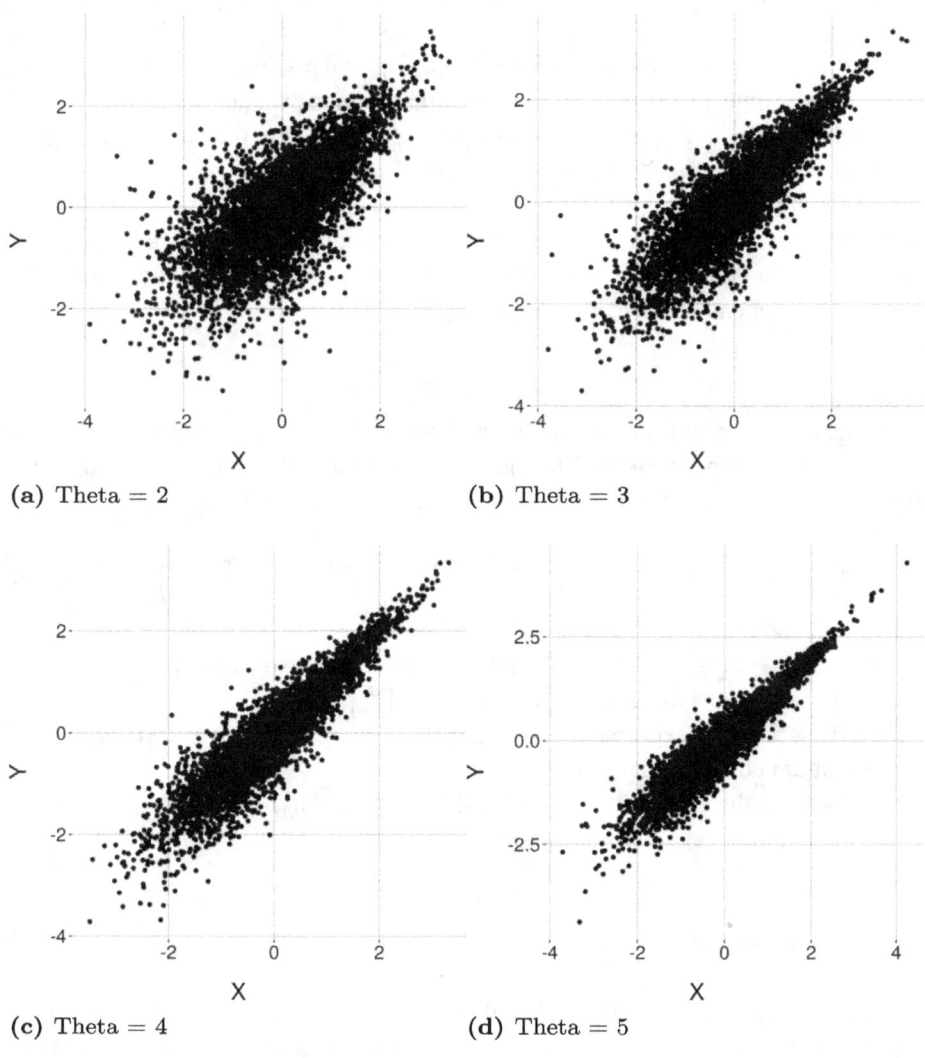

(a) Theta = 2   (b) Theta = 3   (c) Theta = 4   (d) Theta = 5

Figure 14.10: Gumbel copula, bivariate case.

Clayton copula

This is also known as the Cook–Johnson copula. It was initially proposed in [20] and studied later in [77] and [21, 22].

The Clayton copula is given by:

$$C_\theta^{Cl}(u_1, u_2, \ldots, u_n) = (\max(u_1^\theta + \cdots + u_n^\theta - 1), 0)^{1/\theta}, \tag{14.18}$$

with $\Phi_{Cl}(u) = \frac{1}{\theta}(u^{-\theta} - 1)$ (see equation (14.16)) where $u = \{u_1, u_2, \ldots, u_n\}$, is the interval $\theta \in [-1, \infty) \setminus \{0\}$. Notice as well that:

- For $\theta \to 0$ the expression is reduced to the independence copula, $C(u_1, \ldots, u_n) = \prod_{i=1}^{n} u_i$.
- For $\theta = -1$ the expression is reduced to $W(u_1, u_2, \ldots, u_n) = \max(u_1 + u_2 + \cdots + u_n - 1, 0)$, which was indicated before as the maximum negative dependence.
- For $\theta \to +\infty$ the expression is reduced to $M(u_1, u_2, \ldots, u_n) = \min(u_1, u_2, \ldots, u_n)$, corresponding to the maximum positive dependence.

Clayton's copula is an asymmetrical Archimedean copula that exhibits a large dependence at the lower tail [66], as is clear from Figure 14.11. As the association parameter increases this dependence becomes larger.

Frank copula

This copula does not present dependence at the extremes as in the Gumbel and Clayton copulas. It was first proposed in [36] and later studied in [41] and [93], see Figure 14.12 for some examples. This copula is symmetrical and it is specified by:

$$C_\theta^{Fr}(u_1, u_2, \ldots, u_n) = -\frac{1}{\theta} \ln\left(1 + \frac{(e^{-\theta u_1} - 1)(e^{-\theta u_2} - 1)\ldots(e^{-\theta u_n} - 1)}{e^{-\theta} - 1}\right), \tag{14.19}$$

where $\Phi_{Fr}(u) = -\ln \frac{e^{-\theta u} - 1}{e^{-\theta} - 1}$ with $\theta \in \mathbb{R} \setminus \{0\}$ (equation (14.16)). In addition:
- $\theta \to 0$ the expression becomes $C(u_1, \ldots, u_n) = \prod_{i=1}^{n} u_i$.
- For $\theta \to +\infty$ the expression is reduced to $M(u_1, u_2, \ldots, u_n) = \min(u_1, u_2, \ldots, u_n)$ (maximum positive association).
- For $\theta \to -\infty$ the expression is reduced to $W(u_1, u_2, \ldots, u_n) = \max(u_1 + u_2 + \cdots + u_n - 1, 0)$ (maximum negative association).

14.6.2 Simulation with copulas

In many instances, from the available data we are able to infer only the marginal distributions of a collection of random variables. In this case, in order to simulate the joint distribution of the variables, the possibility of using copulas comes in handy. The use of copulas allows us to try different dependence structures, e. g., [66]. In general the steps

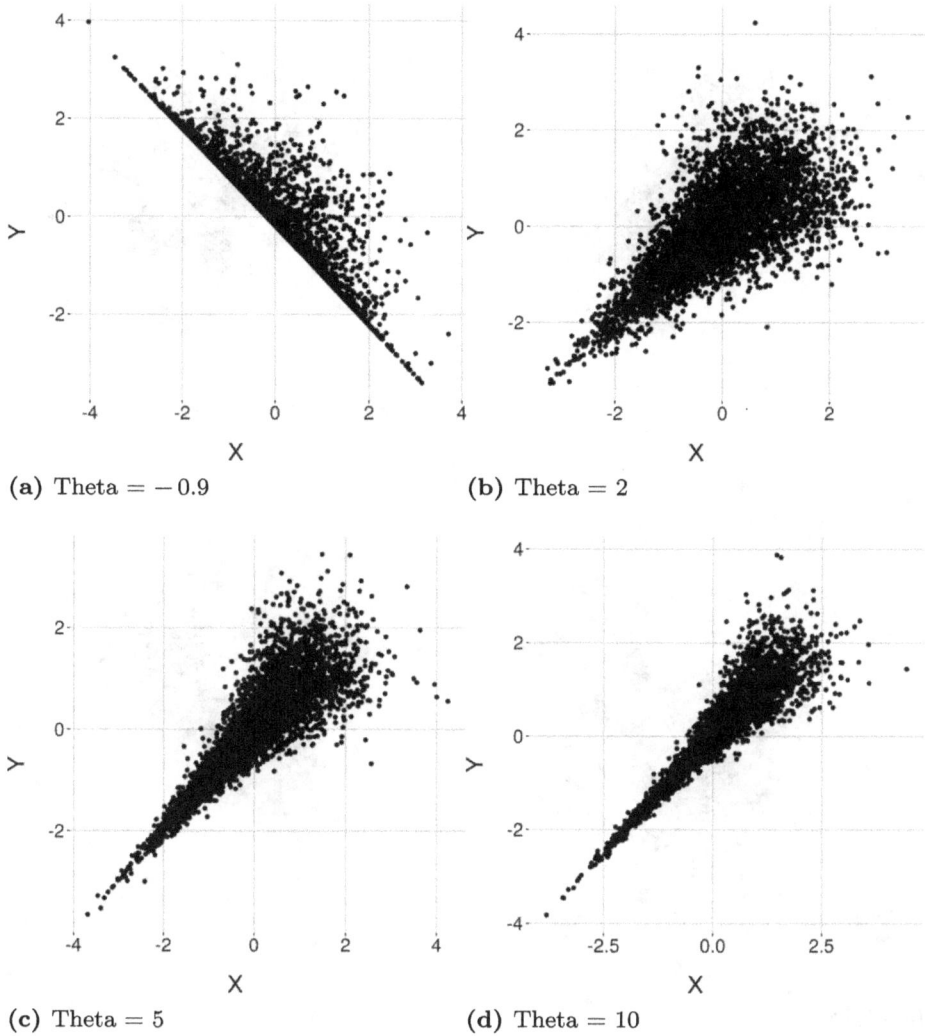

Figure 14.11: Clayton copula, bivariate case.

for the simulation process are:
(1) Estimate marginal distributions $F_1, \ldots, F_n$.
(2) Estimate a pairwise correlation matrix.
(3) Choose a n-copula C.
(4) Simulate random vectors $(U_1, \ldots, U_n)$ from a distribution C.
(5) Apply the transformation $u_i \to F_i^{-1}(U_i), i = 1, \ldots, n$.

To finish, we emphasize that we actually carried out all of these steps in the numerical examples developed to aggregate losses with some form of dependence built in. After

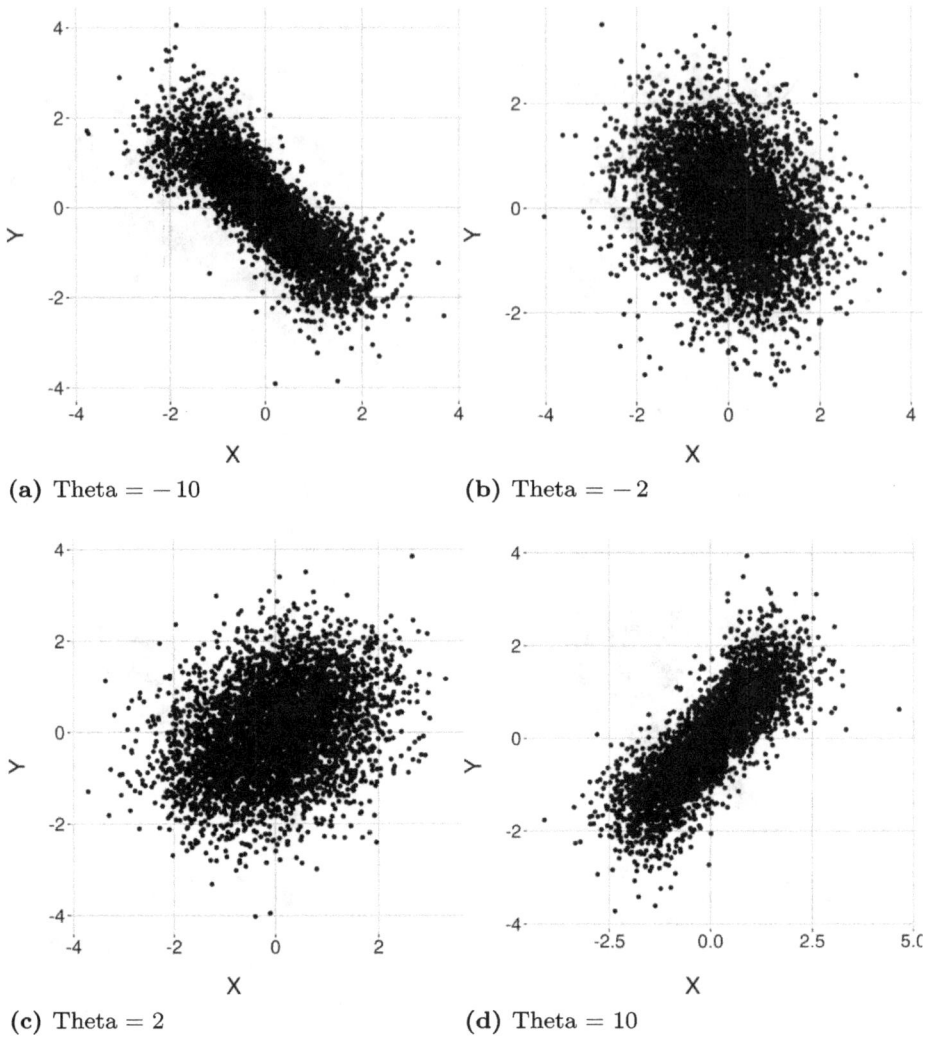

Figure 14.12: Frank copula, bivariate case.

simulating the loss data with the dependence built in, we computed the Laplace transform of the aggregated loss, from which we obtained its density, and then carried out the computations like those exemplified in Chapter 12.

Bibliography

[1] Abbas, A., Maximum entropy distributions with upper and lower bounds, in: Knuth, K. H., Abbas, A. E., Morris, R. D. and Castle, J. P. (eds.) *Bayesian Inference and Maximum Entropy Methods in Science and Engineering*, AIP Conference Proceedings, vol. 803, pp. 25–42, Am. Inst. Phys, Melville, NY, 2005.
[2] Abbott, M. L. and McKinnney, J., *Understanding an Applying Research Design*, John Wiley & Sons, New York, 2013.
[3] Akaike, H., Information theory and an extension of the maximum likelihood principle, in: Petrov, B. N. and Caski, F. (eds.) *Proceedings of the Second International Symposium on Information Theory*, Akademiai Kiado, Budapest, 1973.
[4] Akaike, H., A new look at the statistical model identification, *IEEE Trans. Autom. Control*, **19** (1974), 716–723.
[5] Alexander, C., *Operational risk aggregation*, Operational Risk **4**(4) (2003). http://www.carolalexander.org/publish/download/JournalArticles/PDFs/OpRiskAgg.pdf.
[6] Ang, D., Lund, J. and Stenger, F., Complex variable and regularization methods of inversion of the Laplace transform, *Math. Comput.*, **53** (1989), 589–608.
[7] Antolín, J., Maximum entropy formalism and analytic extrapolation, *J. Math. Phys.*, **31** (1990), 791–797.
[8] Asmussen, S., Jensen, J. L. and Rojas-Nandayapa, L., On the Laplace transform of the lognormal distribution, *Methodol. Comput. Appl. Probab.*, **18** (2016), 441.
[9] Atkins, P. W., *The Second Law*, W. H. Freeman & Co, New York, 1994.
[10] Berkowitz, J., Testing density forecasts, with applications to risk management, *J. Bus. Econ. Stat.*, **19** (2001), 465–474.
[11] Biernacki, C., Celeux, G. and Govaert, G., Assessing a mixture model for clustering with the integrated completed likelihood, *IEEE Trans. Pattern Anal. Mach. Intell.*, **22** (2000), 719–725.
[12] Borwein, J. and Lewis, A., Convergence of best entropy estimates, *SIAM J. Optim.*, **1** (1991), 191–205.
[13] Borwein, J. and Lewis, A., *Convex Analysis and Nonlinear Optimization*, CMS Books in Mathematics, Springer, 2000.
[14] Bozdogan, H., Model selection and Akaike's information criterion (AIC): The general theory and its analytical extensions, *Psychometrika*, **52** (1987), 345–370.
[15] Campbell, J., *The Grammatical Man*, Simon and Schuster, New York, 1982.
[16] Candes, E. and Fernández-Granda, C., Towards a mathematical theory of superresolution, *Commun. Pure Appl. Math.*, **67** (2012), 906–956.
[17] Candes, E., Romberg, J. and Tao, T., Robust uncertainty principles: Exact signal reconstruction from highly incomplete frequency information, *IEEE Trans. Inf. Theory*, **52** (2006), 489–509.
[18] Caussinus, H. and Ruiz-Gazen, A., Exploratory projection pursuit, in: Govaert, G. (ed.) *Data Analysis*, ISTE, London, 2009, doi: 10.1002/9780470611777.ch3.
[19] Christoffersen, P. F., *Elements of Financial Risk Management*, Springer, Berlin, 2012.
[20] Clayton, D. G., A model for association in bivariate life tables and its application in epidemiological studies of familial tendency in chronic disease incidence, *Biometrika*, **65** (1978), 141–151.
[21] Cook, R. D. and Johnson, M. E., A family of distributions for modeling non-elliptically symmetric multivariate data, *J. R. Stat. Soc. B*, **43** (1981), 210–218.
[22] Cook, R. D. and Johnson, M. E., Generalized Burr–Pareto-logistic distributions with applications to a uranium exploration data set, *Technometrics*, **28**(2) (1986), 123–131.
[23] Courant, R. and Hilbert, D., *Methods of Mathematical Physics* vol. 2, J. Wiley Interscience, New York, 1962.
[24] Cover, T. M. and Thomas, J. A., *Elements of Information Theory*, 2nd ed., John Wiley, New York, 2006.
[25] Crump, K. S., Numerical inversion of the Laplace transform using a Fourier series approximation, *J. Assoc. Comput. Mach.*, **23** (1976), 89–96.
[26] Cruz, M. G., *Modeling, Measuring and Hedging Operational Risk*, Wiley, New York, 2002.

[27] de Castro, Y., Gamboa, F., Henrion, D. and Lasserre, J. B., Exact solutions to super resolution on semi-algebraic domains in higher domains, *IEEE Trans. Inf. Theory*, **63** (2015), 621–630.
[28] Dhaene, J., Laeven, R. J. A. and Zhang, Y., Systemic risk: Conditional distortion risk measures, 1901.04689v2, 2019.
[29] Diebold, F. X., Gunther, T. A. and Tay, A. S., Evaluating density forecasts with applications to financial risk management, *Int. Econ. Rev.*, **39** (1998), 863–883.
[30] Doetsch, G., *Introduction to the Theory and Applications of the Laplace Transformation*, Springer, Berlin, 1974.
[31] Donoho, D. and Stark, P., Uncertainty principles and signal recovery, *SIAM J. Appl. Math.*, **49** (1989), 906–931.
[32] Durante, F. and Sempi, C., *Principles of Copula Theory*, CRC Press, Boca Raton, 2016.
[33] Ellis, R., *Entropy, Large Deviations and Statistical Physics*, Springer, New York, 1985.
[34] Feller, W. T., *Introduction to Probability and its Applications*, vol. II, J. Wiley, New York, 1971.
[35] Fontnouvelle, Rosengren E, P. and Jordan, J., *Implications of alternative operational risk modeling techniques*, Technical Report, Federal Reserve Bank of Boston and Fitch Risk, 2005.
[36] Frank, M. J., On the simultaneous associativity of $F(x,y)$ and $x + y - F(x,y)$, *Aequ. Math.*, **19**(1) (1979), 194–226.
[37] Frontini, M. and Tagliani, A., Entropy-convergence in Stieltjes and Hamburger moment problem, *Appl. Math. Comput.*, **88** (1997), 39–51.
[38] Fölmer, H. and Schied, A., *Stochastic Finance*, De Gruyter, Berlin, 2016.
[39] Gamboa, F. and Gassiat, E., Sets of superresolution and maximum entropy in the mean, *SIAM J. Math. Anal.*, **27** (1996), 1129–1152.
[40] Gel, Y. R. and Gastwirth, J. L., A robust modification of the Jarque–Bera test of normality, *Econ. Lett.*, **99**(1) (2008), 30–32.
[41] Genest, C., Frank's family of bivariate distributions, *Biometrika*, **74** (1987), 549–555.
[42] Gneiting, T., Balabdaoui, F. and Raftery, A. E., Probabilistic forecasts, calibration and sharpness, *J. R. Stat. Soc., Ser. B, Stat. Methodol.*, **69** (2007), 243–268.
[43] Gomes-Gonçalves, E. and Gzyl, H., Disentangling Frequency models, *J. Oper. Risk*, **9** (2014), 3–21.
[44] Gomes-Gonçalves, E., Gzyl, H. and Mayoral, S., Maxentropic approach to decompound aggregate risk losses, *Insur. Math. Econ.*, **64** (2015), 326–336.
[45] Gomes-Gonçalves, E., Gzyl, H. and Mayoral, S., Loss data analysis: Analysis of the sample dependence in density reconstructions by maxentropic methods, *Insur. Math. Econ.*, **71** (2016), 145–153.
[46] Gomes-Gonçalves, E., Gzyl, H. and Mayoral, S., Maximum entropy approach to the loss data aggregation problem, *J. Oper. Risk*, **11** (2016), 49–70.
[47] Gomes-Gonçalvez, E., Gzyl, H. and Mayoral, S., Density reconstructions with errors in the data, *Entropy*, **16** (2014), 3257–3272.
[48] Gzyl, H. and Mayoral, S., Determination of risk pricing measures form market prices of risk, *Insur. Math. Econ.*, **43** (2008), 437–443.
[49] Gzyl, H. and Mayoral, S., A numerical approach to the capital allocation problem, *J. Risk*, **23** (2021), 1–24.
[50] Gzyl, H., Novi-Inverardi, P. L. and Tagliani, A., A comparison of numerical approaches to determine the severity of losses, *J. Oper. Risk*, **1** (2013), 3–15.
[51] Hartigan, J. A., *Clustering Algorithms*, Wiley Series in Probability and Mathematical Statistics, vol. 209, Wiley, New York, 1975.
[52] Hartigan, J. A. and Wong, M. A., Algorithm AS 136: A k-means clustering algorithm, *J. R. Stat. Soc., Ser. C, Appl. Stat.*, **28**(1) (1979), 100–108.
[53] Henrici, P., *Applied Computational Complex Analysis*, vol. 2, J. Wiley& Sons, New York, 1977.
[54] Hyndman, R. J. and Koehler, A. B., Another look at measures of forecast accuracy, *Int. J. Forecast.*, **22**(4) (2006), 679–688.
[55] Härdle, W., Kleinow, T. and Stahl, G., *Applied Quantitative Finance*, Springer, Berlin, 2002.

[56] Jakata, O. and Chikobvu, D., Modeling extreme risk of the South African Financial Index (JS80) using generalized Pareto distribution, 2019. https://www.jefjournal.org.za.
[57] Jaynes, E., Information theory and statistical mechanics, *Phys. Rev.*, **106** (1957), 171–197.
[58] Jaynes, E., *Probability Theory: The Logic of Science*, Cambridge Univ. Press, Cambridge, 2002.
[59] Kaas, R., Goovaerts, M., Dhaene, J. and Denuit, M., *Modern Actuarial Theory*, Kluwer Acad. Pubs., Boston/Dordrecht, 2001.
[60] Kiche, J., Ngesa, O. and Orwa, G., On generalized gamma distribution and its application to survival data, *Int. J. Stat. Probab.*, **8** (2019), 85–102.
[61] Kullback, S., *Information Theory and Statistics*, John Wiley, New York, 1959.
[62] Leipnik, R. B., On lognormal random variables: I – The characteristic function, *J. Aust. Math. Soc. Ser. B, Appl. Math.*, **32** (1991), 327–347.
[63] Lewis, A., Superresolution in the Markov moment problem, *J. Math. Anal. Appl.*, **197** (1996), 774–780.
[64] Lin, G. D., Characterizations of distributions via moments, *Sankhya*, **54** (1992), 128–132.
[65] Lin, G. D. and Stoyanov, J., The logarithmic skew-normal distributions are moment indeterminate, *J. Appl. Probab.*, **46** (2009), 909–916.
[66] Lindskog, F., *Modeling dependence with copulas and applications to risk management*, Master's Thesis, Swiss Federal Institute of Technology, Zürich, 2000.
[67] Marsaglia, G. and Marsaglia, J., Evaluating the Anderson–Darling distribution, *J. Stat. Softw.*, **9** (2004), 1–5.
[68] McLachlan, G. J. and Peel, D., Mixtures of factor analyzers, in: Langley, P. (ed.) *Proceedings of the Seventeenth International Conference on Machine Learning*, pp. 599–606, Morgan Kaufmann, San Francisco, 2000.
[69] McNeil, A. J., Frey, R. and Embrechts, P., *Quantitative Risk Management*, Princeton Series in Finance, vol. 10, Princeton, 2005, 4.
[70] Mead, L. R. and Papanicolau, N., Maximum entropy in the moment problem, *J. Math. Phys.*, **25** (1984), 2404–2417.
[71] Mignola, G., Ugoccioni, R. and Cope, E., *Comments on the BCBS proposal for the new standardized approach*, 1607.00756.
[72] Mnatsakanov, R., Hausdorff moment problems: reconstructions of probability densities, *Stat. Probab. Lett.*, **78** (2008), 1869–1877.
[73] Mnatsakanov, R. and Sarkisian, K., A note on recovering the distributions from exponential moments, *Stat. Probab. Lett.*, **219** (2013), 8730–8737.
[74] Moscadelli, M., *The modeling of operational risk: Experience with the data collected by the Basel Committee*, Technical Report, Bank of Italy, 2004.
[75] Nakagawa, T., *Shock and Damage Models in Reliability Theory*, Springer, London, 2007.
[76] Nielsen, R., *An Introduction to Copulas*, Springer, New York, 2006.
[77] Oakes, D., A model for association in bivariate survival data, *J. R. Stat. Soc. B*, **44**(3) (1982), 414–422.
[78] Panjer, H., *Operational Risk: Modeling Analytics*, Wiley, New York, 2006.
[79] Park, M. H. and Kim, J. H. T., Estimating extreme tail risk measures with generalized Pareto distribution, *Comput. Stat. Data Anal.*, **98** (2016), 91–104.
[80] Peña, D. and Prieto, F. J., Cluster identification using projections, *J. Am. Stat. Assoc.*, **96**(456) (2001), 1433–1445.
[81] Peña, D. and Prieto, F. J., Multivariate outlier detection and robust covariance matrix estimation, *Technometrics*, **43**(3) (2001), 286–310.
[82] Ramaswamy, V., Desarbo, W., Reibstein, D. and Robinson, W., An empirical pooling approach for estimating marketing mix elasticities with PIMS data, *Mark. Sci.*, **12** (1993), 103–124.
[83] Raydan, M., On the Barzilai and Borwein choice of step length for the gradient method, *IMA J. Numer. Anal.*, **13** (1993), 321–326.
[84] Rockafellar, R. T. and Uryasev, S., Conditional value at risk for general loss distributions, *J. Bank. Finance*, **26** (2002), 1443–1471.

[85] Ruppert, D., *Statistics and Data Analysis for Financial Engineering*, pp. 183–216, Springer, New York, 2011.
[86] Schwarz, G., Estimating the dimension of a model, *Ann. Stat.*, **6**(2) (1978), 461–464.
[87] Shannon, C. E., A mathematical theory of communication, *Bell Syst. Tech. J.*, **27** (1948), 379–423 and 623–666.
[88] Stephens, M. A., EDF statistics for goodness of fit and some comparisons, *J. Am. Stat. Assoc.*, **69**(347) (1974), 730–737.
[89] Stone, J. V., *Independent component analysis: a tutorial introduction*, Massachusetts Institute of Technology, 2004.
[90] Swait, J. D., Discrete choice theory and modeling, in: *The Oxford Handbook of the Economics of Food Consumption and Policy*, p. 119, Oxford University Press, 2011.
[91] Tagliani, A. and Velásquez, Y., Numerical inversion of the Laplace transform via fractional moments, *Appl. Math. Comput.*, **143** (2003), 99–107.
[92] Tay, A. S. and Wallis, K. F., Density forecasting: A survey, *J. Forecast.*, **19**(4) (2000), 235–254.
[93] Tenney, M. S., Introduction to copulas, in: *Enterprise Risk Management Symposium*, Mathematical Finance Company, Virginia, 2003.
[94] Thas, O., *Comparing Distributions*, Springer, New York, 2010.
[95] Titterington, D. M., Smith, A. F. M. and Markov, U. E., *Statistical Analysis of Finite Mixture Models*, Wiley, Chichester, 1985.
[96] Wang, A. S., Aggregation of correlated risk portfolios: Models and algorithms, *Proc. Casualty Actuar. Soc.*, **85** (1998), 848–939.
[97] Wang, S. S., A class of distortion operators for pricing financial and insurance risks, *J. Risk Insur.*, **67** (2000), 15–36.
[98] Werner, M., *Identification of multivariate outliers in large data sets*, Doctoral dissertation, University of Colorado, Denver, 2003. Available at http://math.ucdenver.edu/graduate/thesis/werner_thesis.pdf.
[99] Widder, D., *An Introduction to Transform Theory*, Acad. Press, New York, 1971.
[100] Willmot, G. E., Panjer, H. H. and Klugman, S. A., *Loss Models: From Data to Decisions*, 4th ed., John Wiley & Sons, New York, 2012.

Index

K-means 181
P-value 15
t-Student copula 196

aggregation of Poisson distributions 6
Anderson–Darling test 190
approximate methods 44
Archimedean copulas 200

barrier crossing times 35
Berkowitz back test 191
beta distribution 24
binomial distribution 18
binomial models 9

chi-squared test 189
claim distribution 30
Clayton copula 202
clustering methods 181
compound random variable 1
convolution of distribution functions 39
Cook–Johnson copula 202
copulas 194
correlograms 191
credit risk losses 31

decompounding losses 136, 146
disentangling frequencies 136

Edgeworth approximation 46
elbow method 187
elliptical copulas 196
EM algorithm 182
entropy function 78
equivalent mixture 147
error measurement 192
exponential distribution 22

fractional moment problem
– with data in ranges 104
– with errors in the data 101
fractional moments 66
Fréchet inequality 195

gamma density 8
gamma distribution 23
Gaussian approximation 45

Gaussian copula 196
generalized moment problem 77
– with data in ranges 102
– with errors in the data 98
generating function 5
goodness of fit test 13, 29
goodness of fit: continuous variables 189
goodness of fit: discrete variables 189
Gumbel copula 200

information criteria approach 187

Jarque–Bera test 191

Kolmogorov–Smirnov test 29, 190

Laplace transform 22, 62
Laplace transform of compound random variables 67
likelihood function 28
lognormal distribution 24
loss aggregation 1

marginal calibration plot 192
maximum copula 195
maximum entropy method 80
maximum likelihood estimation 16, 28, 179
maximum negative association 202
maximum positive association 202
mean absolute error (MAE) 193
method of moments 17, 180
minimum copula 195
mixture of distributions 25
mixture of negative binomials 143
mixture of Poisson distributions 138
moment generating function 17, 22, 40

negative binomial 8
negentropy 188

operational risk losses 30
outliers 186
overfitting 193

Panjer $(a, b, 0)$ class 10
Panjer class 137
Panjer recurrence relations 49

https://doi.org/10.1515/9783111048185-016

Pareto distribution 23
Poisson distribution 6
Poisson mixtures 7
PP-plots 192
projection pursuit technique 186

quantile (Q-Q) plot 28

regulatory capital 134
reliability diagram 192
reliability models 36
root mean squared error (RMSE) 193

sample data dependence 119
sample dependence of the maxentropic density 130
shock and damage models 35
Sklar's theorem 194
standard maximum entropy method 76
standard maximum entropy method extended 97
superresolution 113

time to failure 37
translated gamma approximation 46

www.ingramcontent.com/pod-product-compliance
Lightning Source LLC
Chambersburg PA
CBHW082325220526
45470CB00008B/2408